THE
HIDDEN
UNIVERSE

THE HIDDEN UNIVERSE

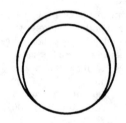

MICHAEL DISNEY

MACMILLAN PUBLISHING COMPANY

New York

Macmillan Publishing Company
866 Third Avenue, New York, N.Y. 10022

Library of Congress Cataloging in Publication Data

Disney, Michael
The hidden universe.

Includes index.
1. Astronomy.   I. Title.
QB43.2.D57 1984   523.1   84-17157
ISBN 0-02-531670-2

Macmillan books are available at special discounts for bulk purchases for sales promotions, premiums, fund-raising, or educational use. Special editions or book excerpts can also be created to specification. For details, contact:
Special Sales Director
Macmillan Publishing Company
866 Third Avenue
New York, New York 10022

10   9   8   7   6   5   4   3   2   1

Printed in the United States of America

For Jan

"I think there is only one way to science—or to philosophy, for that matter: to meet a problem, to see its beauty and fall in love with it; to get married to it, and to live with it happily, till death do ye part—unless you should meet another and even more fascinating problem, or unless, indeed, you should obtain a solution. But even if you do obtain a solution, you may then discover, to your delight, the existence of a whole family of enchanting though perhaps difficult problem children for whose welfare you may work, with a purpose, to the end of your days."

KARL POPPER
Realism and the Aim of Science
Vol. 1, p. 8; Hutchinson 1982

Contents

List of Plates

Acknowledgments

I would like to thank Penny Welch, who typed the manuscript so cheerfully; Malcolm Gerratt, my publisher at Dent's, who waited patiently; Bernard Carr, who advised on two chapters; Huw Jones and Steve Phillipps, who read the whole book and suggested many helpful changes; and my bank manager, for keeping the bailiffs from the door.

Many colleagues have patiently tried to explain their fascinating research to me. If I have misunderstood, oversimplified, neglected or defocussed their work I hope they will forgive me, or at least write or send re/preprints and tell me where I went wrong in time for a second edition, if there is to be one. Missing mass is a vast and rapidly changing field and a snapshot of it from one place at one time can never tell the whole truth.

The author and publishers would like to thank the following for permission to quote extracts or reproduce illustrations from their publications (references in parentheses are to illustration numbers or extracts in the present volume):

The Director of the Lund Observatory (Plate 1).
The Royal Observatory, Edinburgh (Plate 2).
The Carnegie Institute, Washington, D.C. (Plate 3).
The Anglo-Australian Telescope Board (Plates 4, 5 and 6).

Yale University Press (Plate 7, from *The Realm of the Nebulae* by Edwin Hubble).

Dr. N. A. Sharpe, Mount Stromlo Observatory, Canberra, Australia (Fig. 3).

Princeton University Press (Fig. 6, from *Physical Cosmology* by P. J. E. Peebles).

Albert Bosma (Fig. 16, from *Distribution and Kinematics of Neutral Hydrogen in Spiral Galaxies of Various Morphological Types*, University of Groningen).

F. Hohl and *Astrophysical Journal* (Fig. 17, from *Astrophysical Journal*, 168, 343).

André Deutsch and Basic Books, Inc. (extract from *The First Three Minutes* by Steven Weinberg).

The Executors of the late Fritz Zwicky (extract from *Compact Galaxies, Compact Parts of Galaxies, Eruptive and Post-Eruptive Galaxies*, Part 8, p. vi.).

Nature (extracts for reviews in *Nature*, Vol. 280, 1981, 83 and Vol. 294, 1981, 521, Copyright © 1981 Macmillan).

The Quest

MAN is a hunter. His senses were honed to sniff out prey lurking in a thicket and bring it to bay with his spear. Man is an explorer too. Ages before written history little bands of people abandoned their cooking fires and set out on foot to cross mountain ranges and traverse waterless deserts. Others trusted themselves in flimsy rafts to find out what lay round that next bend in the river, or ventured out over the water horizon to discover where the sun set. Long before Magellan or Cook, palaeolithic explorers had reached the windy couloirs of Tierra del Fuego and penetrated the maze-like bush of Tasmania.

But if man was a wanderer he was equally a wonderer too. When they dig up potsherds and axeheads from the past, archaeologists also find monuments and cave-paintings which testify to a mystical inner life, possibly more vivid than the daily round of stone and clay and bone. Man's brain teemed with theories and fantasies. He frightened himself with demons, consoled himself with gods. He lived as much by totem and taboo as he did by root and sinew. Even today, the Aborigine will abandon his government pension and his coke-tins to go walk-about in the dream-time.

Nowadays the animals have turned into circus-performers, into dog-meat and into bangles. The entire globe is mapped and under daily satellite surveillance. And the fantasies? They are in retreat too before the more respectable but equally chimerical abstractions of Marxism, monetarism, political equality and material well-being. But man's instincts, to hunt, to explore and to theorize, remain with him as part of his blood and inheritance. They will no more go away than his fingers will drop off or he will forget how to make love. In answer to the blood-call some men will scale the Eiger wall; others, in envy of the birds, will scheme with bits of metal and cloth until they hang-glide alongside the eagles. Yet others with telescopes and equations will try to fathom the extent, the past history and future fate of the cosmos as a whole. These, the astronomers, are embarked on a hunt, a voyage of discovery, which has really only just begun. What they are trying to do may seem irrelevant at best, preposterous at worst, and arguably doomed to failure in the end. But human history was ever made of forlorn hopes and impossible voyages which triumphed against the odds and in spite of the 'realists' who knew it could not be done.

Many, perhaps most, voyages of discovery begin with a question in the explorer's mind, a question that gnaws away at his curiosity until it becomes an irresistible challenge. The astronomer is now faced with just such a challenge. Within the past few years evidence has been accumulating that there is between ten and a hundred times more mass in each volume of the Universe than we can account for with our present theories and observations. The evidence has been accruing from radio-telescopes, from optical observations, from X-ray satellites, from computer simulations and from mathematical calculations.

Taken separately each testimony is not incontrovertible, but taken together the accumulation of evidence is hard to argue with and even harder to explain. It would appear that despite all our efforts over the centuries we have entirely overlooked the most massive components of the Universe. All round us, pervading the fabric of our Milky Way and far beyond, there is an invisible sea of objects or constituents which comprise ninety percent or more of the mass of our cosmos. And despite our great telescopes and powerful rockets this mysterious agency still maddeningly evades all our efforts at detection. Nevertheless we know it is lurking there because of the effects it has on those structures such as galaxies and clusters of galaxies that we *can* see. Were it not present in great quantities, then spiral galaxies of the sort we live in would fragment into lumps, while the clusters into which the majority of galaxies are clumped would rapidly disperse into open space. Some of the clues have been to hand for a generation or more but only in the last few years has the accumulation of evidence, much of it gathered by accident, put the matter beyond reasonable doubt. We are now faced with the disturbing, and at the same time intriguing, conclusion that most of the fabric of our heavens still remains to be found, and that so far we have merely been scratching at the surface.

Why do we think all this invisible material is present? And if we are right, what form would it take? How might we detect its presence for sure? And if it is present, what effect must it have on the tiny fraction of the Universe at present perceptible to our senses? These questions are fascinating astronomers today. I have written this book, not to answer them all, for that is beyond anyone at present, but to explain how we are trying to answer them and to convey, if I can, a little of the excitement of it all.

To its disciples the first fascination of astronomy lies in great part in its perennial power to set such challenging problems. Even the simplest of observations can lead to profound questions which the mind can hardly ignore.

For instance, if you wander into the garden and look up into the nighttime heavens you will, if it is not too cloudy, see some stars. Ignoring for the moment what the stars are, or how far apart they may be, you will make another observation which is so commonplace that it scarcely seems worthy of remark. The sky is dark. But commonplace as this may seem, the darkness of the sky has some profound implications. It tells us either that the Universe is finite in space, or in age, or in both.

To see why this should be so, consider an analogy. Think of yourself somewhere deep in a limitless forest. But a forest with a difference. All the tree trunks are painted bright orange. Now it does not matter how far away a tree may be from you, its colour will not change. If the forest were truly limitless and you cast a glance from side to side, then in whichever direction you looked your line of sight would intersect an orange tree trunk. In other words, you would be surrounded by a solid wall of orange on all sides, and this would be true irrespective of how far apart the trees were from one another, provided only that the forest continued indefinitely. But if the forest were finite, you would in places be able to see out between the tree trunks and the orangeness would be less pronounced.

Now the forest is a two-dimensional analogue of three-dimensional space, with the orange tree trunks standing in for bright stars. Were the starry heavens to continue indefinitely, no matter how far apart the stars may lie, the eye would in every direction fall upon the burning surface of a star so that the heavens would everywhere glow like a furnace-wall. That the sky is dark, therefore appears to be telling us that the stars cannot continue indefinitely.

Of course you could argue that while the firmament of stars is finite the space surrounding it is not; that it continues indefinitely beyond the last stars. That the stars, in other words, form an island in space.

But this will not work. Stars, we know, exert a gravitational pull upon one another so that an island of stars would fall in upon itself.

The ancients sidestepped this paradox, the so-called Olbers' paradox, by arguing that between the stars there must be some absorbing smoke. By intercepting the radiation from the more distant stars this smoke would save us from being burned alive. But this, as we now realize, is not a valid explanation either. We know today that although energy may be temporarily absorbed it cannot be destroyed. If therefore they absorb starlight, smoke particles must themselves eventually heat up until they too re-radiate like stars. So the paradox remains.

We now recognize only two ways round the difficulty. Either the Universe is finite in spatial extent, or else it has a finite age, in which case there may have been insufficient time for the light from the most distant stars to have reached us. And all this follows simply because the sky is dark at night. We shall argue later on that the finiteness of space by itself is no explanation and that indeed the starry heaven has a finite age.

A second fascination of astronomy, one it shares with many other branches of science, is the way in which an inquiry in one field may lead, by inevitable but unsuspected paths, to questions and speculations in a totally unexpected direction. Thus the commonplace observation that the night-sky is dark leads us to wonder if the Universe may be finite in space. But how can that be? Does finiteness not imply that there must be a barrier at the end of space? But if there is a barrier, surely the wall itself must be part of the Universe, and in any case what lies beyond the barrier?

Sometimes in astronomy it seems as if the answer to one question can be found only at the expense of posing two more that are often a great deal harder to solve; that our small but comfortable island of familiarity is beset on all sides by an endless miasma of perplexing uncertainty. One might be discouraged from setting out on any ambitious investigation by a feeling that it is bound to end in a labyrinthine maze of interlocking paradoxes and seemingly unanswerable questions. After all, is it likely that a creature whose dominion over space and time is so limited, a creature who not so long ago communicated by grunts and cut up his carrion food with a flint, is it likely he could possibly aspire to answer correctly the great questions posed by the cosmos?

The natural, even the sensible response, is to admit that he is most unlikely to succeed. And I must concede that the notion of apes gazing through bits of glass hoping to disentangle the architecture of Infinity does seem at times both comic and preposterous. How, for instance, can we be sure that the same laws of physics, which have been painfully established down here on Earth, are the whole story when one is dealing with incomparably larger volumes of space? Then again the whole of scientific astronomy has at most occupied four thousand years, an interval so short that it corresponds to no more than the eyelid's blink in cosmic terms. This means we can only have what amounts to a single still shot of a Universe which is probably moving and changing decisively in timescales nearer a billion than a million years.

Despite all these quite natural doubts astrophysicists have a great deal of confidence that they can get to the bottom of many cosmic profundities. Their confidence is largely based on the success they have already attained in trying to understand stars. To give us some confidence in tackling our own quest, it is worth looking at how that stellar understanding was won and why we believe it is soundly based.

Goodness knows what speculations must have passed through the minds of early hunters and herdsmen as they gazed up into the bowl of night and wondered about the stars. Were they friendly lanterns hung high above the trees or chinks in the vault of Heaven through which filtered celestial light from a brighter world beyond? The most fundamental question that would have occurred to them, and as it turned out the hardest one to answer, was 'How far away are the stars?' As the techniques of surveying and trigonometry were developed, so they were applied in an attempt to triangulate the height of stars. Unfortunately, such crude theodolites as they had could measure angles with a precision scarcely better than a degree of arc, so it was impossible to say more than that stars were at least 60 miles away.

It would have occurred to some bold spirits to speculate that the stars were other suns, but a very long way off indeed. Such a speculation leads to an estimate of the relative solar and stellar distances as follows.

When you throw a stone into a pond ripples spread outward in perfect circles. When a star shines into space the light waves presumably spread out in all directions, not in circles but in three-dimensional circles, that is to say in spherical surfaces. Now we know the area of any surface depends on the square of its characteristic size. Thus the area of a spherical surface increases with the square of its radius. If the total amount of the starlight is to be preserved as it spreads outwards then the amount of light per unit area of surface must fall off as the square of the distance from the emitting star. When one gauges the apparent brightness of a star by eye one is in effect measuring the starlight which is reaching the pupil *per unit of receiving area* and that is a quantity which falls off with the square of the distance. A star, or indeed any light-source at double the distance, will appear only one quarter as bright. This inverse-square law, as we call it, is quite fundamental to astronomy. Although we cannot test it directly over interstellar distances, we can believe in it so long as we are content that light energy in spreading out is conserved, and that Euclid's laws of geometry apply on a scale of astronomical space.

In practice it was not easy to compare quantatively the relative brightness of the sun and stars when the contrast was so great. Nevertheless it was not too hard to speculate that *if* stars were other suns then they had to be no less than 10^5 (i.e. 100,000) times further away.

This speculation linked the stars to the solar distance and to the relationship of the Earth with the sun, which was another major problem in itself.

The first man brave enough, or foolish enough, to guess that the sun, and not the Earth, formed the focus of the heavenly dance was Aristarchus the Greek. But his hypothesis was rejected, not only on religious and philosophical grounds, but for more logical reasons as well. If the Earth was making its way among the stars, why did not the angles of the stars change relative to one another, as happens to distant objects seen by the moving eye? For instance, if one walks round and round in circles in a wood one notices that the nearer trees appear to change their relative positions compared to those in the background. And the closer the tree the larger the change in angle or parallax as it is called.

Although attempts were made to measure this parallax for the stars as the Earth supposedly moved from one end of its orbit to the other, none was found. When challenged to explain this failure Aristarchus could reply only that the stars must be so far away that the angle was too small to be measurable. In an age when a hard day's journey on horseback was about fifty miles, and when poor people spent their entire lives within a day's walk from home, the implied distances were breathtaking. Not only did man have to abdicate his imagined place at the centre of the Universe, but all his achievements and aspirations would have to be scaled down into comparative insignificance. It is not surprising that Aristarchus' ideas were rejected for 2000 years.

It was not until AD 1600 that Galileo's first telescopic observations of the heavens, and of the planets in particular, forced even sceptics to accede to the heliocentric hypothesis and to the implied enormous distances for even the brightest stars. And it was only in 1838 that Henderson and Bessel, using the precision telescopes made possible by nineteenth-century technology, finally cracked the parallax problem. For even the closest stars the parallax angle turned out to be less than one four-thousandth of a degree, which is the angle subtended by a pea at a distance of about a mile.

So Aristarchus was right after all. Even neighbouring stars are ten million million kilometres apart, a distance so large that it is more conveniently measured in the time it takes light, travelling at 300,000 kilometres a second, to get from one to the next. Whereas the moon is just a light-second or so away, and the sun eight light-minutes, neighbouring stars are typically several light years apart. And applying the inverse square law to the faintest stars seen in a big telescope, some are a million light years away. For starlight to carry so far, the stars could no longer be thought of as the kindly lanterns of God, but must be prodigious furnaces like the sun. The round of speculation, deduction and observation had finally come full circle.

So long as men were able to believe in a very literal interpretation of the book of Genesis, and particularly in the idea that the act of creation had taken place a mere few thousand years before the birth of Christ (4004 years according to Bishop Ussher), there was no good cause to worry how or why the stars might work. Given its enormous mass, which had been roughly estimated by Newton, the sun must have locked up in it adequate heat energy to go on shining for about a million years.

Of all things it was the navvy's spade which was to turn this comfortable conclusion on its head. The layers of gradually evolving fossils which were turned up in canal and railway cuttings spoke of an Earth that had not only a sun to shine upon it but a history measurable in hundreds of millions of years. At the same time, physicists proved that energy was a conserved quantity; that it could neither be created nor destroyed but could only transmute from one form into another.

The crisis came in 1871 when Kelvin and Helmholtz separately proved that no known source of energy could possibly power the sun, or indeed any of the stars, for the long timescales demanded by the geological evidence. How then did the sun shine?

Once parallax became measurable astronomers were in a position to compile a list of the distances to various stars and hence to compute their intrinsic luminosities. In some cases they found stars with luminosities ten thousand or more times greater than the sun's and therefore with a proportionally greater energy crisis.

There then followed a stroke of luck without which we might have remained in ignorance today. Stars are by no means all the same colour. They range from pigeon's blood red through orange, green and yellow like the sun, to white, blue and even violet. And when they compared the colours of nearby stars with their intrinsic luminosities, astronomers found a curious regularity or correlation. The blue hot stars were almost all intrinsically luminous, the red cool stars all, or almost all, intrinsically faint. (See Fig. 1).

This is surprising because the luminosity of a star ought to depend on two quite independent factors: its surface temperature and its surface area or size. We might therefore expect that stars of a given colour (temperature) ought to have a wide range of luminosities depending on their sizes. Not so however. Once the colour is measured the luminosity, and hence by deduction the size, appears to be rigidly determined. It looks as if luminosity, size and temperature are all somehow physically related.

This simple but surprising correlation has a number of significant consequences. To begin with it provides a very simple method of measuring stellar distances. Measure the colour of a star and you can at once read off its instrinsic luminosity from the diagram (Fig. 1). Comparing this intrinsic luminosity with the star's apparent brightness at the telescope yields the distance directly without measuring anything complicated like parallactic angles. The method is fast and reliable out to long range so that at a stroke astronomers were able to survey the three-dimensional geography of the stellar Universe for the first time.

The second consequence of the Hertzsprung-Russell (H-R) correlation, as it was called, is of a more fundamental nature. From it we can infer that stars must be very simple creatures. The colour gives the luminosity and hence the radius. All the gross properties are related. Were stars more complicated, each measurable property could be determined by a wide variety of primary characteristics, so that a simple correlation between them would be unexpected. And this simplicity gives us the confidence to try and understand how they work.

But it gives more than that. It provides a precise test of any possible theory. Any theory which fails to account for the correlation can be rejected out of hand.

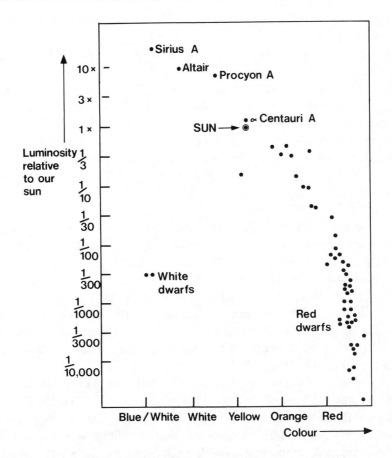

Fig. 1 Hertzsprung-Russell (H-R) diagram. Colours plotted against intrinsic luminosities for the 60 nearest stars, i.e. the ones lying within 16 light years of the sun. This surprising correlation between the two properties—colour and luminosity—was discovered independently by Ejnar Hertzsprung and Henry Norris Russell in 1913, whence the name of the diagram.

Scarcely ten years elapsed between the discovery of the H-R correlation and the emergence of Eddington's theory of stellar structure. A.S. Eddington argued as follows. Judging by the sun, stars are very massive bodies, a million times more massive than the Earth. As such, they are subject to immense gravitational forces which squeeze them inward into a round shape. But if you squeeze ordinary material sufficiently it gets hot. A bicycle pump provides a familiar example. Because the stellar cores are squeezed much harder by the overlying weight they are hotter than the outsides. But heat always flows from a hot region to a cool one, hence heat will be squeezed outward and radiated into space.

The theory is so simple, depending as it does only on the law of gravitation and the simplest ideas of heat flow, that some of his contemporaries, particularly Sir James Jeans, found it hard to take. How, they asked, can you possibly understand a star if you don't know what it's made of inside? 'I don't care what it's made of,' Eddington might have replied. 'Squeezing will render the inside so hot, tens of millions of degrees in fact, that any substance will evaporate into a gas of electrons and nuclei, which is all I need to know. Whether it's made of strawberry jam or dynamite is more or less irrelevant: I will calculate for you what its luminosity must be!'

The sheer effrontery of the theory may have been hard to take, but what could not be denied was its wonderful correspondence with the observations. The main part of the Hertzsprung-Russell diagram was explained while the luminosity of the sun came out as calculated. Moreover, the theory was capable of making all sorts of new predictions which could be and were confirmed by further observations. For instance Eddington's theory predicted that the luminosity of stars should increase with the fourth power of their masses, and this was quickly found to be true.

According to Eddington a star is simply a large lump of matter caught in the relentless vice of self gravitation. Because it is so heavy the interior will be squeezed to a high temperature. Because heat must always flow from a higher to a lower temperature region, the heat will leak outward and so the star must shine. It is the weight of mass which primarily determines the luminosity; the chemical composition of a star is only important in so far as it prescribes how transparent or how well insulated the star will be to its own radiation.

Granted that a star has to shine, we still have to worry about the ultimate source of its energy. Even if, as we now believe, the sun has an internal temperature of about ten million degrees, the total heat reservoir inside it will only supply the outflowing luminosity for a few tens of millions of years. Without some compensating source of energy the interior will cool off, the central pressure will fall, and the sun will shrink.

But the geological evidence shows that far from shrinking the sun has been shining steadily for billions of years.

The mystery was solved during the 1930s by nuclear physicists. They showed that when atoms collide with one another fast enough, fusion will take place, which releases large amounts of energy. The high temperatures in the solar interior imply high velocity collisions between the hydrogen atoms out of which the sun is largely made. Occasionally four hydrogen nuclei will collide and fuse together to form a helium nucleus which will be slightly lighter than the four precursor hydrogens. The discrepant mass does not vanish, it reappears in the form of a prodigious amount of radiant energy via the famous relation $E = mc^2$. In this manner the solar interior will behave like a slowly burning hydrogen bomb which, so long as the hydrogen fuel lasts, can resupply all the energy lost by radiation at the solar surface. Because stars are mostly made of hydrogen (75% by weight) and because hydrogen is an energy-intensive fuel, this hydrogen-burning phase of their lives lasts for billions rather then millions of years.

But the fuel cannot last indefinitely. When it does run out the inexorable process of shrinkage begins once again. The core collapses and the temperatures there rise until helium left over from the initial stage can supply more energy by fusing into carbon. When that is exhausted there is a further shrinkage fusing carbon, by a number of intermediate steps, into iron. One must remember that the shrinkage refers to the main body of the star deep inside. Instabilities may cause temporary expansions and contractions of the light outside layers. This is where the short-lived but spectacular red giant and variable phases can arise.

Once the core has all been turned into iron the star reaches a crisis. Burning iron absorbs rather than produces energy. But the star, for the reasons pointed out already, must continue to shine, and shine prodigiously. Being clamped in an inescapable gravitational vice forces it to radiate.

What happens next is still obscure in its details, but the outcome is certain. The star collapses dramatically. Light stars like the sun collapse into white dwarfs no bigger than the Earth with densities around a million times that of water (see later in this chapter). At such densities electron pressure will support the star and the thermal pressure, which previously gave rise to the high internal temperatures, and hence to the luminosities, can fall away. The luminosity drops right off to a ten thousandth or less of its main sequence value. The star slowly cools through its white dwarf phase to become an invisible black dwarf, composed mainly of crystalline iron.

But only the lighter stars can become white dwarfs. Electron pressure will never suffice to stabilize heavier stars, which must die more violently. When the core of a star twice as heavy as the sun runs out of fuel, it collapses in a matter of seconds into a dense mass of neutrons about ten miles across. During those violent seconds as much gravitational energy is released as the star has radiated in all its previous history. In escaping, this energy blows off the massive outer layers of the star at thousands of miles a second in a titanic explosion which we call a supernova. In a few days the supernova will radiate as much energy as a hundred billion suns. When the ejecta have dispersed as an expanding mass of incandescent gas the dense core in the middle is revealed as a pulsar or neutron star, essentially a gigantic atom nucleus spinning at a 1000 revolutions a second.

For the most massive stars, stars more than five times as heavy as the sun, the initial story is, we believe, the same. But nothing, not even neutron pressure, can hold them up and they shrink *for ever* into oblivion. In accordance with Einstein's theory they become black holes. So dense do they become that the enormous gravitational fields wrap space and time in around themselves. No light, no signals of any sort can escape. They are lost to our ken for ever.

All the refinements of Eddington's original theory have been worked out over the past fifty years. Armed with a more precise knowledge of nuclear physics, and with powerful computers to carry out the complex calculations, we think we can now explain the whole H-R diagram —that is to say, nearly all the stars in the sky. Special triumphs along the way include Chandrasekhar's explanation of white dwarfs, Hoyle and Schwarzschild's calculations of red giant evolution, and Zwicky's prediction of neutron stars forty years before they were actually observed for the first time as pulsars.

As theory has advanced, so has observation. We have now measured the masses, rotation speeds, chemical compositions and magnetic fields of a large number of stars. Variable stars, that is to say stars whose luminosity varies in a regular rhythmic pattern, have attracted notable attention. They vary, we know, because they pulsate rhythmically with periods ranging between hours and months. The pulsations can be mimicked accurately in a computer. It turns out they have a battle going on inside them between radiation pushing its way out and gravitation pulling mass in, neither force being dominant. Because their periods and intrinsic luminosities are closely related, variables are uniquely important as distance indicators. A variable's behaviour makes it conspicuous, and regular observation will establish its period. From the period its intrinsic luminosity can be calculated, and this can be compared with how bright it *looks*. The comparison then allows us to infer its distance away by means of the usual inverse square law. Because stars like Cepheid variables are highly luminous intrinsically (ten thousand times brighter than the sun) they can be used to measure distances out to half a million light years where most other techniques break down.

The theory of stellar structure and evolution today encompasses and explains all sorts of pathological stars as well as the main sequence variety from which it started out. Not only does this theory, started by Eddington, unite a vast number of observations and ideas but it lends us confidence in a much wider sense. Despite his humble origins and puny tools Man has, in a remarkably short time, apparently come to an understanding of the internal workings of something as vast, as inaccessible and as alien as a star. So close is the correspondence between his theoretical predictions and a multitude of observations that it lies beyond the bounds of reasonable probability to imagine the theory is wrong, at least in any fundamental aspect. And if he can tackle the stars what is to prevent him searching, and searching successfully, a great way beyond, out to the galaxies, and to the clusters of nebulae, and so on to the horizons of space and the beginnings of time?

Perhaps the most remarkable thing about stellar space is how empty it is. If we were to shrink the Universe until the sun were no larger than a cherry then the nearest similar cherry or star would be no less than 200 miles away. Not only is the scale hard to imagine but the purpose of creating such an utterly empty Universe seems beyond the grasp of human understanding. More to the point perhaps is whether all this space is really empty. Just because no stars can be seen are we forced to the conclusion that there can be nothing else out there?

Given the vast interstellar voids we must count ourselves fortunate that stars are such luminous furnaces, otherwise we would not see them at all. Had they been a thousand times less luminous, not one, apart from the sun, would be visible to the naked eye. But there is quite a different way of looking at the matter. Exceptionally bright objects are probably very rare and so, like the tallest trees in the forest, they are few and far between. Space could be richly sown with objects less luminous than stars and we simply would not notice them. This places the astronomer in an awkward position. Because distances in the Universe can be so large he can see only the most luminous objects. He may suspect that there is a great deal more dark material around, but as a scientist he cannot believe in what he has no means of detecting. As a working hypothesis he will tend to ignore this dark component of the Universe until, perhaps by indirect means, it forces itself upon his attention. And that is what appears to be happening now. For hundreds of years astronomers have assiduously studied the luminous component of the heavens, building up a picture of the geography and the physics. Only today is it becoming apparent that the bits of the jigsaw will not fit together. More and more we are becoming convinced that the major pieces are still missing.

The study of stellar structure itself points to some important lacunae. Some stars are very short-lived. From what material did they form? Exhausted stars become virtually invisible dwarfs: how many such dwarfs are there? All the stars we can see have masses between fifty times and one tenth of the sun's. What has become of the masses which lie outside this rather narrow range?

We can hope that matter which shines by no visible light nevertheless emits radiation at other wavelengths, perhaps in the infrared, or as X-rays which do not penetrate the atmosphere. A series of spectacular satellite telescopes is now unveiling a new Universe that was previously invisible to the watcher from the ground, and part of the present excitement stems from their findings.

But even if the greatest mass of the Universe emits no radiation at all it will not remain concealed for ever. All matter and energy exerts a gravitational force so that dark material will betray its presence by the gravitational effects it must have on those objects which are bright enough to be seen and studied in detail. Thus it was that the planet Neptune was discovered.

William Herschel, a German by birth, but an Englishman by adoption, was a professional musician. In his spare time, however, he built such fine telescopes and studied the stars so assiduously that he became one of the greatest astronomers of all time. In 1781, using one of his telescopes he discovered by accident a new planet which was to be called Uranus. Uranus, which orbits the sun every 84 years, was found to lie far out from the sun, much beyond Saturn, which explained why it was faint and so had not been found along with the others.

When they measured the progression of Uranus through the constellations, astronomers were disconcerted to find that its orbit was most irregular, not at all in conformity with Kepler's laws for elliptical planetary orbits. In fact no planet moves in a perfect ellipse because each is slightly disturbed by the gravitational pull of the others. Even so, when allowance was made for this in the case of Uranus, there was still a serious discrepancy. Either Newton's gravitational law was wrong or else the errant planet was subject to some extra gravitational influence from a source or sources unknown.

In the 1840s, by which time the anomalies were well established, two mathematicians, quite independently and in ignorance of each other's aim, set out to explain the discrepancies by supposing there to be another undiscovered planet at about twice the distance of Uranus from the sun. One was John Couch Adams, an unknown student at Cambridge. The other was Urban Jean Leverrier, the brilliant young director of the Paris Observatory. Leverrier began work in 1845, but Adams began much earlier in 1841.

It is a fearfully difficult problem to work backward, by hand, from the known perturbations of the orbit, to discover the agency causing them, when you have little idea of its mass or orbit. Nevertheless, on 23 September 1846 a paper on the subject by Leverrier reached the German astronomer Galle at the Berlin Observatory, and the very same night Galle discovered Neptune precisely where Leverrier and Adams had calculated it would be. This triumph of mathematical physics was rightly acclaimed throughout the educated world. For us the discovery of Neptune is significant because it marks the first occasion that a previously unknown body was found through the gravitational effect it exerted on a more luminous and better-studied object. But if it was the first, it was by no means the last and later on a whole new class of very weird and unsuspected stars turned up in the same way.

Being the brightest star in the sky, and therefore probably close by, Sirius was a very natural target for astronomers in search of parallax measurements. The observations were a puzzle, however. Sirius was found to wobble back and forth in its slow progress across the sky. Friedrich Bessel, who made the measurements, attributed this to the pull of a massive but invisible companion star in orbit round it. In 1862 Alvan Clark finally located this companion with his giant new telescope and found it to be 10,000 times less luminous than Sirius, which was puzzling since both appeared to have the same mass. The puzzle was further complicated when Adams found the new star, Sirius B as it was called, was white and therefore hot. To be so hot and yet so faint the star had to be very small indeed, about the same size as the Earth. For so much mass to be crammed into such a small volume Sirius B had to be one million times denser than water. We now know that the star, which was the prototype of many so-called 'white dwarfs' catalogued today, is made out of a previously unsuspected and unfamiliar state of matter, a crystalline gas. White dwarfs are, we believe, the remnants of cores of once luminous stars which have collapsed when their thermonuclear fuel ran out.

Today we can measure the high internal velocities of galaxies, structures much larger than the stars they contain, and we can measure the movements of galaxies relative to one another. In either case can we understand these high velocities if there is no more mass in galaxies than we can attribute to the visible stars they contain. Indeed the visible contents can amount to no more than a small percentage of their total mass. What we aim to do in this book is look at the accumulating evidence for this exciting state of affairs and speculate as to the nature of the mysterious so-called missing mass. If there is enough of it, not only will it affect the dynamics of galaxies, but it will close up the geometry of space and determine the ultimate fate of the universe.

The Stars

BEFORE we go in quest of the missing mass we should briefly review the landscape of the Universe as we see it today. Whereas in the first chapter we sounded an optimistic note, here we must point to some of the outstanding practical difficulties facing those who explore the cosmos. If we recognize how blindfold the astronomer is sometimes forced to go then we shall be all the more alert to the revolutionary discoveries and surprises which certainly lie in wait. Alertness, though, is not enough: we must learn to travel sceptically, for in astronomy even the best-attested 'facts' can rest upon unconscious or shaky assumptions which may fail the test of time. Time and again the maps drawn by one generation of astronomers have had to be redrawn by their successors, and there is no good reason to believe that such revisions have yet come to an end. Certainly if the main constituents of the Universe are still to be found, then we must be prepared for some big surprises. In this chapter and the next I shall be taking the reader on a very brief conducted tour of the Universe as the contemporary astronomer believes it to be. Because we are travelling sceptically we must be concerned not only with what he believes but *why* he believes it.

Understanding the Universe is rather like trying to assemble a giant jigsaw puzzle with no picture of what the end result is likely to be. We do not even know how many pieces the puzzle will finally contain.

Here and there we assemble little islands of neatly interlocking certainty. Unfortunately we rarely know how these islands should lie in relation to one another, while we have even less idea as to the nature and size of the gulfs which could lie between. The real problem is we do not know what we do not know, and in particular we do not know much about what we cannot see.

So what can we see? From our modern towns and cities the answer is 'not very much.' Since the invention of electric light, of street lamps and light pollution, we have turned our eyes away from the starry landscape above. Not so very long ago, as hunters sleeping out in the bush, as fishermen setting night-lines, as herdsmen guarding our flocks, we knew our constellations, christened our stars with names like Dhube and Mintaka, celebrated their rising and setting as harbingers of the seasons, and associated our fates with their own. Nowadays adult intelligent Europeans can be found who have never seen the Milky Way.

To see the heavens in their full splendour it is necessary to travel South, for it so happens that much the more interesting half of the Universe lies above the Southern Hemisphere. And if we are travelling so far we might as well go up into the Andes and see the stars from one of the best astronomical sites in the world. The skies are clear, the mountain air is dry and thin and the atmosphere blowing in over the cool Peruvian current is as steady as it is anywhere on Earth. Come with me then on an observing trip to one of the observatories, which has been built in the Andes in recent years.

The flight out to Santiago de Chile, with intermediate stops in Liberia, Rio, Sao Paulo, Montevideo and Buenos Aires, occupies twenty-seven dehydrating and wearisome hours. Stepping out of the DC-8 into the brilliant Chilean afternoon light one welcomes the sights and sounds of Earth once again, even the hideous Chilean police with their wrap-round dark glasses, Latin moustachios and gold incisors. But the girls are strikingly beautiful, and more than recompense.

The taxi into town weaves it sway through a concourse of battered wheezing lorries loaded down with farm-produce, gas bottles, chicken coops, humanity, livestock, all in a bustling hurry to get somewhere else. The land on either side reminds me of Spain, with water melons and maize growing. The Andes brood in the background. Nearer the city we pass through a populous shanty-town of lean-to shacks, suicidal chickens and stately women carrying their water home in sawn-off oil cans.

We pass by forbidding blackstone Spanish churches and through market squares braying with donkeys to the centre of old Santiago, which I find remarkable for the profusion of its brilliant flowers and for the extremes of poverty and riches one sees standing side by side on the same pavement.

The Observatory has thoughtfully provided a brief hostelry of rest for us travel-fatigued observers. Drawing up in a leafy cul-de-sac the driver leaps out to tinkle a bell in what appears to be a monastery wall. A silent maid in white glides before me to show the way to my room. It is old and Spanish, opening through French windows to gardens on either side. Beside the bed there is a salver of drinks, an epergne of fruits, flowers everywhere. But after the flight, it is the silence I treasure, and the drowsy afternoon peace, broken only by the gushing of fountains.

Entranced almost, I step through the curtains out onto the lawn. The garden is mine alone; only the bees come to share the solitude and sip at the fountain. Wandering among the trellises of passion-flowers and bougainvillea I discover a secret swimming-pool. Glancing furtively about I strip off and plunge in. Diving through the bubbling green I have time to reflect that astronomy has its consolations.

To tea, which is serviced in the shade of the lemon-trees, come other astronomers, as weary as I, just off their planes from Stockholm, Frankfurt or Paris. Astronomy being a village, we have all met somewhere before, perhaps at a mountain observatory, maybe at a conference in Sicily. And like astronomers anywhere, we forget our surroundings, idyllic as they may be, to talk shop.

Our idyll is all too short-lived, for early next morning we are off in a station-wagon for the 600-km drive north along the coast to the Observatory. The pacific will be on our left the whole way, the Andes to our right. Northern Chile is the narrow strip jammed between the two.

The views can be breathtaking, for in many places the road switchbacks along the sea-cliffs, as others plunges down to cross an Andean valley. The ocean itself is mostly shrouded in low-lying fog caused by the freezing Peruvian current, a current which hugs this coast, rules the whole climate and economy of the hinterland, and lends to Chile its chief attraction for astronomers. As they cross the current the warm damp Pacific Westerlies are cooled, their moisture precipitates out in fog-banks over the sea, and as the airs rise over the Andes there is no moisture left to form cloud or rain.

No moisture means virtually no vegetation and therefore few people. But at the bridge of Punta de Aconcagua we get out to stretch ourselves amidst the bright crimson and blue tents of a gypsy encampment.

These, and a few incredibly hardy families who make a living working tiny seams of copper and smelting the ore themselves, are almost the only people we see.

The last hundred kilometres or so appears as desolate as the moon. Even the cactus has given up, for here in the Atacama desert it may not rain more than once in a decade. Although the air is clear enough to see a hundred miles, there is nothing to see apart from the Andean cordillera itself which broods overhead like an immense tidal wave of stone. Up on its face lie the telescopes of the European Southern Observatory, which is to be our destination.

After the hard grind up we find ourselves on a terrace overlooking the desert with the clouded Pacific beyond. The untidy huts straggle amongst a waste of sun-shattered boulders. Only the glaring silver and white telescope domes remind us that this is not a construction camp. But at seven thousand feet we notice that the sky is not light blue, as it is at ground level, but cobalt blue with the hue of space itself. During the day all is somnolence and quiet, but at dusk the astronomers wake for breakfast and the Observatory comes to life.

Since my first night is a period for acclimatization and readjustment before the serious business begins, I can take my time over coffee watching the sun set red into the Pacific fogs. The stars come out one by one but I bide my time before going outside to savour a view I have been looking forward to for months.

No one who has not seen for themselves can imagine the true wonder of the Southern Heavens seen from such a site. Stars blaze and twinkle right down to the horizon; one can almost hear the distant rumble of the Universe. The enormous archway of pale light which stretches far overhead is the Milky Way as it can never be seen from the North. Strands of light wander off among the constellations, interrupted by dark holes and clouds. Even the brightest stars seem to be gathered towards it. Overhead in Sagittarius there is a great bulge of illumination. Altogether it is a spectacle which excites more than wonder; it poses questions which demand answers.

What are stars, which to the naked eye are merely points of light? What more can we learn by inspecting their images through a telescope at high magnification? I have a key to one of the lesser telescopes not being used tonight, so we can go and take a look.

The reflecting telescope is basically a simple instrument which has not changed in its essentials since Newton designed the first one more than 300 years ago. Ours has a 40-centimetre diameter curved mirror which collects and focusses the light at the eye-piece. This last serves to provide the required magnification, something we can easily change by swapping the various eye-pieces in and out. The telescope tube is mounted on a clockwork drive which compensates for the Earth's rotation, so keeping a target star in the field of view. The dome around it, with a slit which can be opened up, serves to protect the instrument from the sun and elements, and to shield the telescope from the wind at night.

Since the sky has been marked off by celestial latitude and longitude I can set the telescope up on my desired star simply by looking up its celestial coordinates in an almanac. If I set the machine to look at Sirius all I can see through the low magnification eye-piece is a brilliant point of light, brighter by a million times than I can see with the eye alone. While I can discern no shape I can at least see that Sirius has a creamy white colour, distinctly lacking the familiar yellow tinge of the sun.

At low light levels the retina loses its colour sensitivity, which is why to the naked eye, most stars look the same bluish-white. Through the telescope, though, the full colours are restored. If we move from Sirius to Mintaka in the nearby constellation of Orion, we find a star that is sapphire blue. Moving the telescope again to Betelgeuse in the same constellation we see a reddish orange star tinged like the embers of a dying fire. Redder still with the hue of ruby is Mira the demon star. Blue, white, yellow, orange and red, seen through even a small telescope like this one, each bright star has a distinctive colour of its own, and these colours can be accurately measured by comparing the light to be seen through different coloured filters.

An even more effective way of distinguishing the colours is to remove the telescope eye-piece and replace it with a small spectrograph. This is nothing more than a prism with an eye-piece of its own. If we focus on Sirius once again we see not a single point of white light but a line of brilliantly coloured images, each formed of a spectrally pure band of light. To the left the images are ultraviolet and as we move across the spectrum to the right we see images that are successively blue, green, yellow, orange, red, crimson and deep infrared. Sirius normally appears white because the images in all hues are of roughly equal strength. If we move back to Betelgeuse we see the spectrum has the same line of rainbow-colours but now we notice that the blue and green images are fainter while the orange and red ones are much enhanced. Conversely Mintaka's light is piled up to the violet-blue end of the spectrum, and tails away noticeably towards the red. Every star is found to have a distinctive spectrum, and these spectra can be photographed and examined in minutest detail.

What do these different colours or spectra signify? We might guess they have something to do with the stellar surface termperatures because our experience of fires, furnaces and welding arcs tells us that the hotter the flame the more blue/white the colour will be.

We can check this guess by returning to Sirius and examining its spectrum more carefully. If we do so we shall find half a dozen or so slight, but distinct breaks in the spectrum, places or colours in which Sirius appears to be emitting no light at all. Now physics tells us that light of different colours simply corresponds to light-waves of different wavelengths and it is possible to measure the wavelengths of the absorption breaks, or 'absorption lines' as they are called, rather precisely. Once the spectrum is mapped out like this in some detail one can compare the results with the spectra of various chemicals that have been heated to a high temperature in the laboratory. The spectrum of Sirius turns out to be that of a high-pressure mixture of hydrogen and helium gas admitted with traces of other chemicals like oxygen, nitrogen and carbon, heated to a temperature of 10,000°C. The spectrum of the red star Betelgeuse, when studied in sufficient detail, is found to be quite different, with literally hundreds of absorption lines in it. Nevertheless its spectrum can also be reproduced in the laboratory by heating roughly the same mixture of gases, but to only 3600°C this time. Stellar spectra, it transpires, are highly sensitive to temperature, allowing us to confirm with great certainty that the colours of stars are indeed a measure of their surface temperatures. The rare blue stars like Mintaka have surface temperatures of 50,000°C or more, yellow stars like the sun are at about 6000°C, while the coolest red stars have temperatures as low as 3500°C or less. Remember, though, that these temperatures refer only to the shallowest surface layers from which light can escape directly. The great bulk of stellar interiors must be much hotter in order to sustain the pressures necessary to support the stars against their own gravity.

Stellar spectra betray far more about stars than their temperatures. We have seen already that the absorption lines tell us what stars are made of. They are composed of a mixture of elements or atoms heated to the gaseous state, for solids or liquids would simply evaporate. The atoms concerned are those we are familiar with here on Earth, but the exact mixture is very different. Stars are made predominantly out of the light gases hydrogen (75% by weight) and helium (24%), with all the remaining elements, principally oxygen, nitrogen and carbon, making up less than 1%. With a few rare exceptions most stellar compositions turn out to be much the same, though in some very old stars the elements heavier than helium are much depleted. The light elements hydrogen and helium, so common in the stars, and indeed throughout the cosmos, have presumably been largely lost to the Earth because its gravity is too weak to retain them.

A stellar spectrum contains literally thousands of pieces of information which astrophysicists have spent the past century learning how to decode. We can routinely tell the size, rotation-speed, surface gravity, surface pressure, chemical composition, magnetic-field and line-of-sight velocity of a star from its spectrum. For instance, suppose we had found the distance to Betelgeuse (590 light years) by some other techniques. From the distance and the apparent brightness we can use the inverse square law (Chapter 1) to infer the star's intrinsic luminosity, which turns out to be no less than 16,000 times greater than the sun's. But the spectrum informs us that the surface temperature is only 3400°C. Our physical theories, confirmed in the laboratory, tell us that a surface at 3400°C will radiate no more than 760 watts per square centimetre. In order to radiate like 16,000 suns it follows that Betelgeuse must be of enormous size, a red giant of a star so large that if it were placed at the centre of the solar system the Earth would be orbiting deep inside it. I am assuming of course that Betelgeuse is spherical in shape like our sun. But can we see the shape of stars if we look at them through a telescope?

If I greatly increase the magnification by changing the eye-piece the point image of Sirius is replaced by a wobbly jelly-fish of a thing which wanders restlessly about, which expands and contracts and appears to flicker in brightness. We have run into the so-called 'seeing' and no amount of extra magnification will help at all.

Because the 'seeing' differs between two telescopes looking simultaneously at the same star it cannot be a property intrinsic to the star itself and we now recognize it to be an effect of the atmosphere through which we are forced to look. When we notice the road surface in front of us shimmering on a hot day we are observing 'seeing' operating on a different scale. Just as water bends light so does air, but by different amounts depending on its temperature. Microthermal fluctuations on the scale of a few centimetres, either round the telescope or in the atmostphere above it, distort all celestial images, setting many fundamental limitations to the accuracy of observations we can make from the ground. For instance, no matter how large our telescope, we cannot hope to see features on the moon less than 2 kilometres across or on the sun less than 600 kilometres. We are not able, nor do we expect, to see the real shape of even the closest star. Because of atmospheric blurring we see the Universe much as a fish at the bottom of a pond sees the outside world through the ruffled surface of the water. Not only does 'seeing' prevent us from perceiving the shapes of stars directly, it prevents us from measuring their brightnesses, their positions and their movements with the precision we would like to have. There are subtler handicaps too. Nowhere is the sky absolutely dark. By introducing an obstruction into the telescope field it is easy to see that the obstructed field is much darker even than the darkest patches between the stars. This means that in smearing out the image the seeing will lower the contrast between a faint star and the background light against which we have to view it. Thus seeing limits the faintest stars we can detect and so limits the horizons we can hope to reach.

Bad as it is, seeing could have been much worse. Aristarchus pointed out 2500 years ago that if the Earth goes round the sun, then as seen from the Earth the nearby stars ought to describe ellipical orbits against the background of stars much further away. He showed, by geometry, that the angular sizes or parallaxes of these ellipses, if they could be measured, would yield the distances to the stars describing them.

Although they tried, the ancients were unable to detect any such parallactic motion whatsoever. They were forced to conclude either that the stars were enormously distant, or else that Aristarchus' conjecture about the Earth travelling around the sun was wrong. For historical reasons the ancients judged against Aristarchus and the heliocentric hypothesis was lost for two thousand years.

When Galileo's telescopic observations showed, beyond any reasonable doubt, that the Earth does indeed orbit the sun, then the search for stellar parallaxes, and thus for the distances to stars, began in earnest. The parallaxes turned out to be so small, the distances implied so great, that it was to be 1837 before Friedrich Wilhelm Bessel measured a parallax for the star 61 Cygni of 0.3 seconds of arc. Now one second of arc is one three-thousand-six-hundredth of a degree, the angle subtended by a pea 1 kilometre away from you. More crucially, it represents just about the smallest angular scale the atmospheric seeing will allow the astronomer to measure. Only by combining statistically the results of many, many observations was Bessel able to obtain the first reliable interstellar distance. Had the seeing been but three times worse, then we might still be ignorant as to these distances, and to so much of the astrophysics that follow from them. One may well wonder how many more discoveries of similar significance are still veiled from our sight by the atmosphere. For instance we cannot know whether planets orbit any sun but our own.

Nothing much could really be known until the stellar distance problem was unlocked. When Bessel in Germany, Henderson in South Africa, and Struve in Russia independently turned the key at the same time, astronomers were rightly ecstatic. Some flavour of their excitement is conveyed by the Victorian prosiness of Sir John Herschel, then our greatest astronomer, when he presented a medal to Bessel:

> Gentlemen of the Royal Astronomical Society, I congratulate you and myself that we have lived to see the great and hitherto impassable barrier to our excursions into the sidereal universe, that barrier against which we have chafed so long and so vainly . . . almost simultaneously overleaped at three different points. It is the greatest and most glorious triumph which practical astronomy has ever witnessed.

So by the lucky chance that the seeing-size of stars is one, rather than three, seconds of arc, man was able to squeeze between the atmospheric bars and reach out for the Universe. It was to prove an awesome experience.

As a first step astronomers, measuring the distances to some of the brighter stars, were able to calculate their intrinsic properties and so compare them with the sun. Table 2.1 (see page 28) lists the intrinsic properties of the ten apparently brightest stars in the sky obtained in this way. It turns out that far from being uniquely bright the sun is dwarfed by some of its better-known companies. Look at Rigel, the bright blue star in Orion. Its luminosity, no less than 44,000 times greater than the sun, is almost terrifying. We can be thankful that Riget and the enormous red supergiant Betelgeuse are such a long way off.

So where does the sun fit into the scheme of things? Judging from Table 2.1 the sun is not only insignificant, it is the faintest star. And that seems an unlikely coincidence. Indeed there are one or two suspicious things about Table 2.1. Take the two supergiants Rigel and Betelgeuse. Not only are they much brighter than the rest, they also happen to be much further away. Indeed they are so far away that we would not see them at all unless they were supergiants.

The problem with Table 2.1 is that it contains stars selected in the first place precisely because they appear to be bright, and so it should come as no surprise to find that most if not all of them are intrinsically bright as well. For that reason they cannot form a typical sample of stars in general, and it is hardly fair to compare the sun with them. A more typical sample of stars ought to be those lying within a fixed distance of us, say within 16 light years of the sun. Because stars can wander freely in and out of it, the solar neighbourhood ought, at any one time, to contain a typical sample of the stellar population. But how do we pick out these nearest neighbour stars, particularly if many of them turn out to be faint and otherwise insignificant? The trick is to use the apparent stellar motions.

Stars do not stand still: typically they drift randomly about at speeds of around 20 kilometres a second in relation to their neighbours. 20 kilometres a second is about 50,000 mph, which may seem a lot, but to set it in perspective we have to think of the enormous sizes of stars and the unimaginable distances between them. Travelling at 20 kilometres a second the sun would need 8 hours to cover a distance equal to its own diameter. To reach its nearest neighbour star, Proxima Centauri, would take 60,000 years. If we return to Earth 100,000 years hence, we should notice that the constellations had changed a bit, but most of them would still be recognizable.

In a year a star will move about a billion kilometres. The angular distance it will be seen to move, as judged from the Earth, will depend on its distance away from us. If it is relatively very close, i.e. about 5 light years away, its angular motion, or 'proper motion' as we call it, will amount to about 2 arc seconds, which is just measurable. If it is ten times further away the angular motion is only 0.2 arc seconds, which is imperceptible. This suggests a clever way of picking out our neighbour stars. We look for all the stars in the sky with the relatively higher proper motions. Amongst these should be all our closest neighbours. Thus it was that Bessel picked out 61 Cygni as a suitable candidate for his pioneering parallax determination.

It is of course quite impossible for astronomers to measure the positions and motions of all the stars in the sky every year. But it is feasible, using a wide-angle telescope fitted with a camera, to photograph the whole heavens every decade or so. Plates of the same region of sky taken ten years apart can then be compared, essentially by lying one plate on top of the other. The vast majority of distant stars will be found not to have moved. But our few nearest neighbours will be conspicuous as having jumped by small but noticeable amounts from one plate to the next. This 'proper motion' technique has been used to identify all the nearest stars, and so to assemble what we believe is a typical sample of the overall stellar population. And it is this sample against which we should more fairly compare the sun. Data for the ten nearest stars to us, discovered in this way, are set out in Table 2.2.

Contrast Table 2.2 with Table 2.1. To begin with, notice that with the exception of Sirius and the Southern double star α Cen (with its two components A and B) all our closest neighbours are inconspicious objects appearing many thousands of times less bright than the prima donnas of Table 1. Were it not for their high proper motions (column 6) we would not have known they were nearby at all. The reason they look so faint is that indeed they are intrinsically very feeble radiators (column 4), with luminosities hundreds or even thousands of times lower than the sun's. For instance, if the Earth were in orbit around Wolf 359 then that star's feeble red glow would cast no more light on us than does our moon. Ninety-five per cent of the light in the solar neighbourhood is given off by Sirius A alone, with the sun providing most of the rest. So while the sun is no prodigy it does at least come second in a class composed mostly of very dull red dwarfs. Also note that one star, the white dwarf Sirius B, is much smaller in radius than the rest, being no larger than the Earth. Another surprise is to find that single stars are in the minority, six of the ten being double-stars or binaries in orbit around one another. As we shall see later, these binaries hold the clue as to the masses or weight of stars.

Table 2.2, which contains the most representative sample of the stellar population that we know of, could not have been assembled without the aid of photography. Indeed so much of our astronomical knowledge is due to photography that we should say something of its powers, and its limitations.

Photography was introduced into astronomy about a century ago, since when it has brought a revolution greater even than the invention of the telescope. There are three reasons for this. First, by exposing our films or plates for hours at a time we can detect objects well beyond the reach of the human eye. Second, the developed plate provides a permanent and objective record from which quantitative measurements, the basis of any science, can be made. Third, by fitting cameras to wide-field telescopes we can study large areas of the sky and literally tens of thousands of stars and nebulae at once. And not only the images of the stars can be recorded; we can photograph their spectra as well. It is no accident that within fifty years of taking up photography astronomers had discovered how stars work, plumbed the immensities of intergalactic space, and surprised everyone by showing that the Universe expands.

So much of our present knowledge is founded on photography that we would be wise to be conscious of its very real limitations, the chief of which is the extremely limited storage capacity of a photograph per unit area. No part of the sky is completely dark: from even the emptiest hole in it we receive a feeble glow, partly from scattered man-made

TABLE 2.1

Ten apparently brightest stars

Star	Apparent brightness rel. to Sirius	Distance light years	Luminosity rel. to sun	Colour	Surface temp. °C	Radius rel. to sun	Remarks
Sirius	1.0	9	19	White	9,500	2.5	Brightest apparent star
Canopus	0.5	200	5,200	Yellow/White	7,000	60	Southern star
Rigil Kent	0.36	4.2	1.4	Yellow	5,700	1	Southern double star
Arcturus	0.30	36	83	Orange	4,400	16	Orange giant
Vega	0.28	27	44	White	9,900	2.5	White main sequence star
Capella	0.25	46	120	Yellow/Orange	4,900	16?	Yellow giant
Rigel	0.25	880	44,000	Blue/White	14,000	35	Blue supergiant
Procyon	0.19	11	58	Yellow/White	6,000	12	Yellow subgiant
Betelgeuse	0.19	590	16,000	Red	3,400	800	Red supergiant
Achernar	0.17	110	525	Blue/White	16,000	5	Blue subgiant
Sun			1	Yellow	5,700	1	Yellow main sequence

TABLE 2.2

Ten nearest stars

1	2	3	4	5	6	7	8
Star	Apparent brightness rel. to Sirius	Distance light years	Intrinsic luminosity rel. to sun	Colour	Proper motion sec/year	Radius rel. to sun	Remarks
Proxima Cen.	1/75,000	4.3	1/14,000	Ruby	3.9	1/3	Red dwarf
α Cen A	1/4	4.3	1.3	Yellow	3.7	1	Double main sequence
α Cen B	1/15	4.3	1/3	Orange	3.7	0.7	Stars
Barnards star	1/25,000	5.9	1/2,500	Ruby	10.3	1/3	Red dwarf
Wolf 359	1/1,200,000	7.6	1/70,000	Ruby	4.7	1/8	Red dwarf
Lal 21 185	1/4,000	8.2	1/190	Red	4.8	1/2	Red dwarf
Sirius A	1	8.7	19	White	1.3	2.5	Double main sequence
Sirius B	1/12,000	8.7	1/500	White	1.3	1/40	White dwarf
L726–8A	1/350,000	8.7	1/17,000	Ruby	3.4	1/3	Double red dwarf
L726–8 B	1/600,000	8.7	1/30,000	Ruby	3.4	1/3	

lights, partly from the atmosphere, partly from sunlight reflected off interplanetary dust, partly from a host of background stars and nebulae too feeble to be seen as individuals. The result is that after about an hour's exposure a plate begins to blacken all over, and we have to stop. And putting the plate into a big telescope does not help, though it may shorten the exposure, because the telescope amplifies the unwanted background light as well as the stellar images. If we combine this maximum 3600 sec exposure with the fact that photography is only about one per cent efficient (i.e. of every 100 photons that fall on a plate, 99 go undetected) we must realize that all we have to go on is a 3600/100 or a 36-second look at the cosmos. Despite the undoubted achievements of photography surely we must be very wary of a picture of the Universe gleaned from a single thirty-second glance!

Returning to our zoo of stars we now know that while the vast majority of the populace consists of feeble red dwarfs, most of the light is provided by luminaries like Sirius and supergiants like Rigel. But we are chiefly interested here in the masses of the stars. Is the bulk of cosmic material to be found amongst the myriad faint dwarfs, or is it locked up in the rare but super-luminous giants?

The mass of a celestial body can be inferred from the gravitational pull it exerts on its satellites. By comparing the pull required to hold the Earth in its orbit with the gravitational force measured in the laboratory between two lead balls, Michell and Cavendish found that the sun weighs 2,000,000,000,000,000,000,000,000,000 tons. Put in another way, the sun is a globe roughly a million miles in diameter with an average density equivalent to that of water. Alternatively, one can say that the sun weighs 2500 tons for every kilowatt (1 electric fire-bar) of heat and light it emits. Despite its gaseous constitution the sun is a truly massive body, about half a million times as massive as the Earth.

Unhappily even the nearest stars are too far away for us to see their planetary satellites, if indeed they possess any. Fortunately, however, about half of all stars are double or multiple, and we can sometimes observe the orbital motions of the components about one another, and so infer the gravitational forces, and hence the masses at work.

If I train the telescope on α Cen, the second closest star to us (Table 2.2), then when I look through the eye-piece I see that α Cen consists of a pair of close, but distinctly separate stars. The brighter one is yellow like the sun, the orangy-red companion is almost four times fainter. They appear to be roughly ten seeing-diameters apart.

If I came back to reobserve α Cen tomorrow night I would find no apparent change in the separation and orientation of the two companions. But if I come back in ten years time a noticeable movement will have taken place. By the patient accumulation and measurement of such observations, astronomers have found that the two stars in a α Cen orbit one another once every 80 years, and that at maximum separation they are about 35 seconds of arc apart. Assuming that the laws of gravitation are the same out there as they are here, we can calculate that the yellow star is 8 per cent heavier than the sun, the orange one 12 per cent lighter.Since the yellow star is 30 per cent brighter than the sun and the orange one 60 per cent fainter, it looks as if the luminosity of stars increases quite sharply with their masses. And this inference is born out by the studies that are now complete for literally hundreds of visual pairs. For non-giant stars, at least, the luminosity is found to increase with the fourth power of the mass. Thus a star only ten times heavier than the sun will be a supergiant no less than 10,000 times more luminous. Conversely the common red dwarfs that are ten thousand times fainter intrinsically than the sun nevertheless have masses that are ten percent as great. All this is neatly explained by Eddington's theory of stellar constitution (see Chapter 1).

Of course it is a nuisance to have to wait a century for important measurements like these. In any case the method only works on stars that are close enough to us to be resolved in spite of the atmospheric seeing. There is, however, quite another way of detecting and weighing binaries, which is independent of the separation or seeing, requiring only spectroscopy.

If we train the telescope on Spica we see a bright blue star which does not look double at all. But if we look at it through the spectrograph we notice something odd: all the absorption lines are doubled. And if we were to follow it for several nights we would find that the two sets of lines separate and close back together again in a rhythmic way, with a cyclic period of 4 days. Spica consists of two blue giants orbiting so closely around one another that they cannot be separated by eye. The movement of the lines is brought about by the famous 'Doppler effect' as the two stars change their high orbital velocities in the line of sight direction.

Since the Doppler effect is one of our most useful astronomical tools we should briefly explain how it works. Imagine a friend who is throwing tennis balls towards you at what he counts to be one second intervals. If he stands still, the balls will reach you one second apart. But if he moves towards you each ball has a slightly shorter journey to travel than its predecessor, and so the balls will reach you rather less than a second apart. Conversely, if he is moving away, the balls will arrive at longer than one second intervals. Now particular types of atoms in a star emit or absorb light waves with a precisely determined frequency or colour, that is to say the wave crests leave the atoms at precisely determined intervals. But if the star is approaching the observer then, by the same argument that applied to the tennis-balls, the intervals will appear to be shorter, that is to say the apparent frequency will rise. In other words, the light appears to be bluer or 'blueshifted' by an amount depending upon the star's speed of approach, or 'redshifted' if it is a receding. By comparing the stellar spectrum lines with lines produced by atoms at rest in the laboratory, the astronomer can measure velocities of approach and recession along the line of sight with great precision.

Spica then is a 'spectroscopic binary' as opposed to a visual binary like α Cen. It consists of two blue giants each about 8 times as massive and a thousand times as luminous as the sun, rotating round one another at hundreds of miles a second.

Binary observations have built up for us a systematic picture of stellar weights. The most massive supergiants are about fifty times heavier than the sun; the faintest dwarfs only one tenth as heavy. We believe that stars outside this rather narrow mass image simply would not work. If they are too heavy, radiation pressure renders them unstable and they either collapse or explode. If they are too light they never generate internal temperatures sufficient to get nuclear reactions under way, so that they cool out of sight.

Although giants are massive they are very rare, and the only reason we notice them is that they can be seen a long way off. The lighter dwarfs are very much commoner, but inconspicuous. A careful census then shows that overall there are about equal amounts of cosmic material locked up in giants, in solar-type stars, in red dwarfs and in white dwarf stars like Sirius B. It follows that none of the very different kinds of stars can be ignored in any serious inventory of the visible cosmic mass.

One of the reasons why giants appear to be so rare is that they are, we believe, comparatively short-lived prodigally burning up fuel to produce their spectacular luminosities. But how can one speak of the age or evolution of a star when in cosmic terms the human life cycle and indeed the whole span of human history is but the blinking of an eye? One could stare at a typical star for 10,000 years and nothing would change about it. Since they do not perceptibly age it is impossible, by studying them as individuals, to answer questions about the age and evolution of stars.

Anyone who observes the heavens for a few moments, however, will notice that many stars appear to occur in groups. Some of these groupings, and indeed most of the constellations, are merely fortuitous alignments of prominent foreground and background stars. But some of the tighter groupings mark real clusters of stars bound to one another gravitationally and moving together through space. The Pleiades, sometimes known as the Seven Sisters, is a prominent cluster of this type to be seen in the North. Because the stars within a given cluster presumably have a common origin, and therefore a common age, clusters can be used to explore the behaviour of stars in time.

Let me set the telescope on Hyades, the closest such cluster to us. You can see the brighter stars of the cluster with the naked eye, they live in the horns of Taurus the Bull. If I fit the lowest power eye-piece, because I am interested not in large magnification by large field of view, and look through it, then I see a crowd of over a hundred brightish stars scattered across the whole field with a background of many more fainter stars lying in between at the limit of visibility.

The Hyades are too far away for a parallax determination of their distance to be accurate. But being a bound group they are all moving through space together in the same direction, and the 'convergent point'—i.e. the point in the sky to which they all appear to be travelling—can be measured with the aid of pairs of proper-motion plates taken a few years apart. Knowing this point, and the Doppler velocity of the stars, it turns out that we can calculate the Hyades distances as about 130 light years. Knowing the distances, we can calculate the luminosities and it turns out that of the 350 stars in the cluster there are only a few giants and no supergiants. One also notices that most of the brightest stars are either white or yellow with the largest number of fainter stars being orange. What has become of the giants and supergiants and why are there no blue stars at all?

The obvious next step is to look at other clusters, of which there are many hundreds accessible with the telescope. Some few like the Pleiades do contain blue stars and blue giants but in most, as in the Hyades, such stars are absent. Indeed even white stars are missing from many clusters whose population consists entirely of cool stars as cool as or cooler than the sun.

The explanation for their colour differences is not hard to find. Blue giants are prodigious wastrels. They burn so brightly that their fuel cannot last for long. Presumably clusters like the Pleiades which still contain blue giants are young, while clusters like the Hyades are so old that the blue stars have all burned out. This picture is confirmed by computer calculations based on Eddington's theory of stellar structure. A comparison of the hottest stars in any cluster, with the computer calculations, then allows us to determine that cluster's age. Young clusters like the Pleiades are only a million or so years old, whereas the Hyades are 400 times more ancient. But even the Hyades are young compared with some old red clusters which have been around for 10 billion years or almost since the Universe began. Dating the structure of the cosmos by this intercomparison of cluster observation and complex calculations has been a triumph of modern astrophysics, and would probably have been impossible without the electronic computer.

The study of clusters in the 1930s also brought to light the existence of material in between the stars. Distant clusters were often found to be fainter and redder than they ought to be if compared with those nearby. Trumpler attributed this to the existence of absorbing smoke in interstellar space. This was confirmed by spectroscopy which showed that some stars have sharp calcium and sodium absorption lines in their spectra with redshifts that are incommensurate with the redshifts of the normal stellar lines, suggesting they arise in gas or smoke travelling at a different velocity, but lying in the foreground line of sight to the stars.

The existence of this absorbing smoke immensely complicates the astronomer's task. The smoke blocks out much of the Universe from sight and inconveniently reddens the stars we can see (notice the reddened sun at dusk on a murky day), making distance determinations even more difficult than they are already. On the other hand the existence of this interstellar material does suggest the likely origin for the young clusters we discussed above. We now know that young stars are invariably found in regions choked with smoky gaseous materials, and the presumption is that new stars condense out of this diffuse gas. Calculations show that individual stars cannot condense on their own; only whole clouds or clusters at a time. We believe that even a lone star like the sun was born in such a cluster containing hundreds or thousands of companions, but that in the course of time the cluster either dissolved by the evaporation of its members, or else was broken up by close collisions with other large structures.

In making a mass inventory of space it is obviously important to know how much material is locked up in the gas and smoke, or 'interstellar matter' as it is sometimes called. Fortunately even a cold gas strongly emits radiowaves at certain wavelengths characteristic of its composition and temperature, and one of the achievements of radio astronomy has been the mapping of the gas in our neighbourhood, and far beyond. Most of it turns out to be very cold hydrogen (about 10 degrees above absolute zero) in either atomic or molecular forms, confirming the supposition that hydrogen is the primitive 'stuff' out of which all else is made. Although it is more of a nuisance, the smoke, which is believed to consist largely of soot and fine sand, comprises only about one per cent by weight of the interstellar gas. Some of the smoke originates, we believe, in low temperature stars which burn like celestial bonfires, the rest is probably the debris from supergiants exploding at the end of their lives. The total weight of gas near the sun is no more than a third of the weight locked up in visible stars.

We have said something about how stars form when interstellar gas clouds collapse, of how and why they shine while their hydrogen fuel lasts, and of the narrow mass range in which stable stars can exist. But we are left with a number of pertinent questions about stellar evolution in general. For instance, what happens to stars when they run out of fuel, as clearly some of the more luminous ones must do in timescales much shorter than the age of the Universe? And what becomes of condensing fragments of matter that are either too massive or too light to form stable stars? We have only incomplete answers to these questions, partly based on observational evidence, and partly the result of theoretical calculations.

Consider first a star twice as massive as the sun. Both observations and theory tell us that it will be fifty times more luminous than the sun. With fifty times the output but only twice the fuel the star can last for only a per cent or so of the age of the cosmos (about 10 billion years; see later).

When the hydrogen runs out the star does not suddenly stop shining —it cannot. The interior has to remain hotter than the outside simply to support itself, and the temperature gradient thus set up necessarily entails an outward flow of heat. But with no fuel to resupply the lost energy, the interior, by now mostly composed of helium left over from the hydrogen-burning stage, must contract under the inexorable force of gravity. But as the interior density and gravity rise so must the pressure, and hence the temperature. The temperature gradient becomes even steeper and so ironically the star is forced to become even more luminous that it was before. A vicious circle develops. The luminosity rises, the core contracts and grows hotter, the luminosity rises further still until the star is several hundred times more luminous than the sun.

To allow all the extra luminosity to escape, the outer stellar envelope swells up dramatically until the star becomes one of the enormous red giants seen in clusters of the appropriate age. During this phase the atmosphere is no longer stable and considerable mass may be lost into space by so called stellar winds.

Such a situation cannot continue for long because energy must be found to replace what is being radiated. There is temporary relief when the internal temperature reaches about 10^8°C, at which point collision between the helium atoms is sufficient for them to fuse into carbon, releasing further thermonuclear energy.

Unfortunately the star has become by now so luminous that the helium, which is in any case a rather thin fuel, cannot last for long. What happens next is by no means clear in detail, for the vital stages are probably too short-lived to be easily observed. What is certain is that the core must collapse rapidly to a density about a million times higher than water, at which point the star finds permanent relief. Relief comes about because the electrons, long stripped from their atomic nuclei, are now so closely packed together that they repel one another by a weird quantum-mechanical effect called degeneracy pressure. This pressure, which will support the star for all time henceforth, derives from the density of the material, and not its temperature, and so the core can cool off at last. With the central temperature falling the heat-gradients which drive the luminosity fall away; the star dies and shrivels up to become a white dwarf like Sirius B. Such an object, no larger in size than the earth, but a million times as massive, can generate no further energy, and so it will cool to total obscurity, taking a billion years to do so. Younger white dwarfs can be dimly seen and several are known in the solar neighbourhood, marking the tombstones of once spectacular giants. In a billion years or so the sun will follow the same course, becoming in the end a white dwarf less than a thousandth as luminous as it is now. White dwarfs form the end-point of evolution for stars in the range 0.5 to 2 solar masses. Although they are dim, they are both numerous and heavy, representing a significant reservoir of cosmic mass.

If we next consider a star of 5 solar masses the initial course of evolution is not dissimilar. The starting luminosity, though, is now 2000 times solar, so the fuel runs out in about a million years. However, when the core contracts in such a giant, electron degeneracy pressure provides insufficient relief, and there is a violent central collapse in a matter of seconds. Such enormous temperatures are reached that atomic nuclei are first torn to bits and then turned back entirely into neutrons. The core can only stabilize when the neutrons are crushed cheek by jowl to the incredible density of a thousand million tons per teaspoonful. The gravitational energy released in the collapse exceeds all the luminosity radiated during the previous million years. Thus a detonating shockwave rushes outward blowing the outer envelope of the star off at speeds of ten thousand kilometres a second or so. We witness what is called a supernova, an exploding star which can radiate for a while (a few weeks) as much energy as ten billion suns. Such spectacular events are of course hard to miss, and one takes place in the Milky Way every century or so. For a while such detonating stars can be seen even in broad daylight. The supernovae of AD 1054 and 1652 were indeed so violent that they frightened the populace and found their way into historical documents. The remnants of such explosions can be identified today as clouds of incandescent gas still expanding at enormous speeds into space.

Most of the material in a 5 solar mass star is thus returned to the interstellar medium as gas which can be recycled into further generations of stars. The gas, though, has now been largely cooked or processed into heavier elements such as oxygen, carbon and iron, and it is likely that the atoms in our own bodies originated in such cataclysms which took place before the sun and Earth were formed.

The energy of such an explosion has to be paid for somewhere. It was calculated forty years ago that the core should remain as a gigantic atomic nucleus about ten miles across. It was thought that such a neutron star, as it was dubbed, would be far too small and inert to be detected. But in 1967 Jocelyn Bell discovered sharp pulses of radio emission emanating from several places in the Milky Way. They were called pulsars. Further research showed that the pulses could only originate in stars spinning so rapidly, in one case at several hundred times a second, that in order to hold themselves together against centrifugal forces the stars could be no more than a few miles across. And when one was found in the heart of the supernova explosion of 1054, known now as the Crab nebula, there could be no room for doubt. Neutron stars, so long predicted by theory, had been found at last. Why they emit radio pulses remains a mystery; it probably has to do with the fact they are enormously strong spinning magnets. At all events the radio output derives from the energy of rotation, because the spin rate is observed to slow down gradually. After a few million years the rotation drops to less than once per second, at which point the radio pulses cease, and the neutron star fades into undetectability.

The fate of even more massive stars, and stars 80 times as heavy as the sun are known, is understood with much less certainty. They can

be observed as blue supergiants in their youth, but they are so prodigal
of energy they cannot last for long, which explains why they are so
rare in space, and why they are never seen in clusters more than a few
million years old. As hydrogen runs out the core must collapse, but
now it is too massive to stabilize either as a white dwarf or even as a
neutron star. Possibly what happens is that the collapse continues for
ever. For ever and into oblivion. In accordance with Einstein's theory
it would become a black hole. So dense do such phenomena become
that enormous gravitational fields are set up which wrap space and
time around them completely so that no light, no signals of any sort
can escape. They are lost to our ken for ever.

Whether black holes really exist, and how many of them there are,
remains for now a topic of active debate. Satellite telescopes have
recently located binary stars which emit copious X-rays varying violently in timescales of a millisecond. The visible star of the pair, which
itself cannot be the source of the X-rays, is found from its Doppler
shifts to be orbiting a massive but invisible companion which is certainly responsible for the X-rays. This X-ray companion must be extremely dense to exhibit such rapid variability, but is found to be too
massive to be either a neutron star or a black hole. By a process of
elimination we are led to speculate that a black hole is present. We
mark its presence only because it is so close to its visible companion
that it is sucking off its envelope, which is then swallowed by the black
hole, emitting a last scream of X-rays as it disappears. The observations
all fit and it is certainly hard to think of an alternative explanation,
though not everyone is convinced. Of course, because of their very
nature it will always be difficult to prove the existence of black holes,
but the evidence is building up. Certainly we have to reckon with the
possiblity that all massive stars rapidly become black holes, which, in
the absence of close companions, will be totally invisible.

Light stars, in the range between a half and a twentieth of a solar
mass, burn so slowly as feeble dwarfs that they continue almost indefinitely. They are observed to be very numerous in space, and therefore
are an important component of cosmic mass. But with the faintest
detected specimens only a millionth as luminous as our sun, it would
be easy to underestimate how many there could be.

Condensations twenty times or more lighter than the sun can support
themselves by degeneracy pressure, without the necessity for high internal temperatures and the consequent outpouring of radiation. Such
objects, of which Jupiter may be considered a specimen, are sometimes
called black or brown dwarfs. Very little is known about them apart

from the fact that they could embrace a lot of mass without making themselves known to us.

The whole course of stellar evolution, as we know it today, is crudely summarized in Fig. 2. Large clouds of interstellar gas collapse and fragment into stars of different weights. The stars evolve according to their masses, giving off radiation, returning some of their substance to the interstellar medium, and leaving behind various invisible remnants. The net effect is the transmutation of detectable gas into undetectable superdense objects like black and white dwarfs, black holes and neutron stars. The process is irreversible because escaping radiation has left the superdense objects with enormous energy debts they can never repay. We cannot avoid the conclusion that much of the cosmic mass may be hidden in dark exotic objects whose presence will

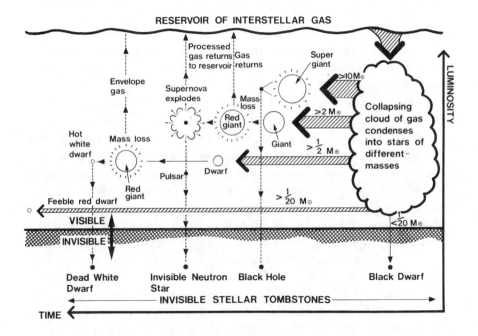

Fig. 2 Stellar evolution crudely illustrated, with stellar luminosities portrayed upwards and time towards the left. On the right a large cloud of interstellar gas is condensing into stars of different weights ($>10M_\odot$ denotes a star with more than ten times the sun's mass). The rest of the diagram shows what happens to these stars during their evolution. The net effect is that stars return some of their material as gas to the interstellar voids for further recycling, while the rest finishes up in the dense invisible bodies shown at the bottom. Only red dwarfs evolve too slowly for much to have happened to them during the history of the Universe so far.

only be detectable through the gravitational influence they exert on luminous stars nearby.

It will turn out that what we are most interested in is not so much the mass of the cosmos as its mean density, that is to say the mass per unit volume. But volume measurements imply distance determinations which are never easy to make in astronomy. Indeed as we go outward beyond the immediate neighbourhood direct parallaxes are too small to be measured accurately and we must rely on a series of successively less direct and less reliable distance-surveying tools, until our best estimates are uncertain by a factor of two and sometimes worse. Indeed, around 1950 it was discovered that all large-scale distances had been underestimated by a factor of no less than 10, and violent arguments about the subject are still going on. It is worth reviewing the distance problem if only to keep in mind the uncertainties involved.

Atmospheric seeing renders direct parallax measurements inaccurate beyond 100 light years, and even within that range it is such a troublesome technique that only a couple of thousand parallaxes have been measured since Bessel's first in 1837.

For greater distances, right out to the horizon, the favoured method is the 'standard candle' technique. One seeks to find objects or 'standard candles' in the far distance that are in all respects similar to an object nearby whose distance has already been found. Comparing apparent brightnesses of the far and near candles, and applying the inverse square law, it is then possible to calculate their relative distances. There are, however, several snags.

Suppose we wanted the distance to a very faint cluster in which something interesting, say a supernovae explosion, was taking place. What we do is to look for some star in the cluster with clearly recognizable characteristics. The favourite candidates are pulsating variables, that is to say stars whose luminosity varies rhythmically as they expand and contract with periods of between hours and weeks. Such variables can easily be picked out from the rest by comparing a series of photographic plates. Moreover studies have shown that variables with the same period are usually identical in their other properties, and in particular in their luminosities. This should not surprise us because stars are basically so simple, and we understand the pulsation as a temporary battle between gravitation and pressure going on in the star's atmosphere during a specific phase of their evolution.

Suppose in our faint cluster we find a yellow star pulsating regularly every 15 days. Such a star is called a Cepheid variable. To find the distance all we then have to do is to measure our Cepheid's average apparent brightness and compare it with the brightness of another 10-day Cepheid which is close enough by to have had its distance measured reliably by some other techniques, for instance by parallax.

Here enters the first difficulty. Because their pulsational phase is short-lived, Cepheids are very rare and there is none so close to us that its distance has been reliably determined. Since we cannot use parallaxes we resort to other less reliable techniques. For instance to get at Cepheid luminosities Harlow Shapley invented in ingenious statistical method. He measured the proper motions and radial (Doppler) velocities of the few nearest (i.e. brightest) Cepheids he could find. He then argued that these stars ought to be moving about randomly in space, and that therefore their radial velocities ought to be on *average* equal to their motions transverse to the line of sight. But transverse motion equals measured angular proper motion multiplied by the unknown distance. So he computed an average distance for his Cepheids so as to make the two measured components of velocity equal. Once he had their distances and brightnesses he could compute their absolute luminosities. He found that Cepheids with a period of 15 days are intrinsically ten thousand times as luminous as the sun. Because they are so luminous Cepheids make ideal standard candles because we can pick them out a long way off.

Comparing Shapley's value for the luminosity with the measured brightness of our own 15-day Cepheid we compute a distance for it of, say, 2000 light years.

But there are complications. Can we be sure that our faint cluster Cepheid is really identical to the nearer-by Cepheids in Shapley's survey? Could not differences in age and composition render the comparison invalid, or at least inaccurate? More seriously we have to worry about obscuration along the line of sight. Along a distance of 2000 light years surely there must be a great deal of obscuring smoke, smoke which would make a naive use of the inverse square law inapplicable? And then again what about Shapley's Cepheids? They are hardly likely to be free of absorption either. Did he compensate sufficiently?

I do not want to give the impression that distance determinations are impossible. They are not: for instance additional very careful spectroscopic and colour measurements can be used to reduce some of the above uncertainties. Moreover clever people are always coming up with refinements and with new and ingenious techniques. So far, however, all of them have their difficulties and qualifications, and most depend on a sequence of assumptions which cannot be checked with the precision we would like. So we must not lose sight of the fact that we suffer from myopia. Distances in the Universe are so immense by comparison with the size of our solar system that we do not have an adequate base-line for triangulations. In consequence we must be cautious in interpreting observations which sometimes depend on an uncertain 'guesstimate' of a large distance.

Each generation of astronomers tries to build a revised, and hopefully more faithful, picture of the Universe out of such imperfect observations as it has been able to make. The picture must change, sometimes radically, because the observations, which are neither as precise nor as complete as we really need, are improving all the time. The seeds of each review are usually to be found in the imperfections of the last, or in the inadequacies of the observations upon which they were based. So it would be well to summarize at least those weaknesses of which we are presently aware.

First there is the atmospheric 'seeing' which severely limits the amount of detail we can perceive and which in particular renders the direct measurement of distances to all but the nearest stars impossible. We have partially overcome the problem by indirect means, by the use of standard candles such as variable stars, but these are fraught with uncertainties which mount until in the far distance our guesstimates are accurate to no better than a factor of two or so.

Second is the enormous disparity between the human time-frame and the cosmic pulse-rate. A single turn of our Milky Way occupies five times as long as all of evolution since the dinosaurs. There has not been, nor will there be, time enough for us to see things really happen. We must work with what amounts to a single still picture of the grand dynamic scheme. It is as if we had to deduce not only the rules of a soccer game but the final score as well, when all we have to go upon is a still photo of some instant of the game.

Third we must content with the observations and reddening caused by huge clouds of interstellar smoke which veils large parts of the cosmos from sight. Fortunately modern radio and infrared telescopes are beginning to probe these previously hidden regions.

Fourth, our present picture of the Universe is largely based on photography of the sky. While the photograph, in its ability to integrate over time, far surpasses the eye as a detector, it nevertheless quickly saturates, so that observations so far amount to no more than a brief half minute peek at the cosmos.

Finally our most severe constraint is illustrated by comparing the information in Tables 2.1 and 2.2 earlier in this chapter. Our observations naturally focus in on the bright stars that can be easily seen but which in fact are very rare and in no way representative of stars as a whole. This natural tendency to pick out the spectacular as opposed to the truly prevalent is called 'observational selection,' and it has lied to us, and will lie to us time and again. We have seen the struggle that was involved in locating our nearest neighbour stars in space, by the method of proper motion. For the most part they turned out to be unremarkable objects previously hidden away against a cloud of background stars. But at least they are stars that shine, with nuclear furnaces in their cores. But what of the myriad objects out there, which may dominate the mass density of space, yet which may shine by no light of their own: comets, asteroids, planets, black dwarfs stars, black holes, rocks, pebbles, dust and a host of exotic sub-atomic particles like neutrinos and magnetic monopoles, to say nothing of unimagined and unimaginable denizens of the deep and dark?

There is one hope of finding out. All matter and energy, no matter how dark and invisible, exerts a gravitational force, or so we believe. If there is something out there, there is a hope of finding it: by the gravitational influence it must exert on the visible population of space. Let us therefore step outside our telescope dome for a while, and out onto the Chilean mountainside once again. The moon should be down by now, enabling us to see some of the grand design of the Universe.

3

The Milky Way and Beyond

WE have learned something of stars as individuals; it is now time to find out how they are aggregated in space. The darkness of the night sky already tells us they cannot continue indefinitely in all directions.

Stepping outside the telescope dome we see at once why it was worth coming all this way. The moon has set and our eyes have now become adjusted to the dark. There are no city illuminations to distract us here; indeed as we look down into the desert not a single light can be seen. By contrast, overhead the firmament of stars burns and twinkles as you have never seen it before. That pale archway of light that vaults from horizon to horizon, culminating in the great bulge of Sagittarius, is the Southern Milky Way. Far down in the South we see two islands of pale light, resembling fragments of the Milky Way that have broken off and drifted into space: those are the Large and Small Magellanic clouds, named after the great navigator. Within the Milky Way itself one discerns large black holes and striations which we might now guess to be clouds of absorbing smoke, as indeed they turn out to be.

We know that the Milky Way continues around the Northern Hemisphere so that a belt of nebulous light entirely girdles the Earth. The ancients speculated as to what it was made of, but lacking the means to find out they settled for the name *galaktos* denoting 'milk' in Greek.

If we step back into our telescope and focus it on any part of the Milky Way, we see in the eye-piece a teeming myriad of faint stars like germs on a microscopic slide. Photographs of the Milky Way, which go a hundred times fainter than the eye (Plate 2), show star clouds so crowded that the stellar images appear to overlap one another. There are millions, so many countless millions of stars, that we are clearly looking at a structure of gigantic size.

Some idea of the scale can be inferred from Table 2.1, a compendium of the brightest apparent stars. Presumably when we look at regions a long way off it will be the brightest stars in that region that we shall notice once again. Judging from Table 2.1, a typical bright star has an intrinsic luminosity of about 500 suns and stands a distance of a 100 light years away (these are very rough averages). Now a comparison of the Milky Way stars in our eye-piece with those stars in Table 2.1 shows that the Milky Way stars are apparently ten thousand times fainter. Applying the inverse square law leads to the inference that they are $\sqrt{10,000}$ or a hundred times as far away. In other words our crude guesstimate suggests that the portion of the Milky Way we can see is about 100×100 light years or 10,000 light years away. And if we multiply up the number of stars which we can see in the belt per square degree by the area of the belt, then the Milky Way is found to contain ten million or more perceptible stars. But as we have found out before, the perceptible stars are the very rare but luminous giants, so that if we take account of the dwarfs as well there ought to be at least a billion stars in the Milky Way.

We lie then in a stupendous structure at least 10,000 light years across. But if we count the relative numbers of bright and faint stars in a direction at right angles to the belt, we find the thickness of the belt to be no more than a few hundred light years. It looks as if we dwell in an enormous slab or disc of stars about fifty times as wide as it is thick, a sort of island Universe in Space. Outside the disc the density of stars falls away rapidly, perhaps to nothing, and this would provide an explanation for the Olbers' paradox.

Two questions now spring to mind. First, why should such an enormous edifice apparently centre itself on an insignificant star like the sun? Second, why does it not collapse on itself through the force of gravitation? The answers to both can be found in the movements of the stars.

Proper motion studies, combined with Doppler velocity measurements, enable us to build up a picture of stellar motions in the local neighbourhood. Study of the data leads to a remarkable discovery. One finds the stellar motions are highly systematic. Stars which are neighbours in space move in a well-regulated order rather like fish in a shoal. There is a general migratory pattern. The neighbouring stars are moving along the plane of the Milky Way, at right angles to the great bulge overhead in Sagittarius. They appear to be swinging in a great procession around the bulge, as if they were in a circular orbit around it which will take 250 million years to complete. Since the sun is, we believe, about 5000 million years old, we have been round and round that bulge about twenty times already.

But why should the bulge be singled out as the centre of motion? After all it is not so much brighter than the rest of the Milky Way. The answer must be that the bulge is very massive indeed, so acting as the centre of a sort of gigantic solar system. Observations of stars in the bulge show it to be about 30,000 light years away. Combining this distance with the measured rotational speed of the stars, about 250 kilometres a second, and applying the law of gravitation, we find that the bulge must weigh as much as a hundred billion suns. We are part of an enormous rotating stellar system called the Galaxy, centred on the bulge in Sagittarius. It is a sort of stupendous rotating discus, with Sagittarius at the centre, the sun and its neighbours lying far out somewhere near the edge.

This dynamic picture does not seem to square with the obvious appearance of the Milky Way, which looks more like a roughly uniform flat slab, with us at centre. But the discrepancy can be explained by the manifest existence of large clouds of obscuring material lying in the galactic plane. This smoke largely shrouds the centre of the Galaxy from sight and indeed obscures most of the structure except a smallish region of the outer disc (Plate 2).

This is such a startling picture that we should not be too eager to take it on trust. Incontrovertible support, however, has come in the last thirty years from radio telescopes. Radio waves are unaffected by smoke, penetrating it freely. The cold hydrogen gas, found everywhere between the stars in the Milky Way, copiously emits radiation at a wavelength of 21 cm. With radio telescopes we can map out the whole structure, including the radial velocities. There can be no doubt about it. The Milky Way is an enormous fried egg composed of stars, interspersed with smoke and gas. The sun is somewhere out near the edge, and the Sagittarius bulge is the central yolk. The whole structure, which we call the Galaxy, rotates about the bulge once every quarter billion years. It is like another solar system but on a much grander scale. Instead of ten planets we have a hundred billion stars all swirling in a celestial dance controlled by gravitation. From where we are, smoke obscures all but the local details. What a pity. What a spectacle it would make, could we but see it all.

We can now understand why the Galaxy does not collapse upon itself through self-gravitation. The spinning motion generates a centrifugal force which counterbalances gravity just as it prevents the Earth from falling into the sun.

Now that we know about the Galaxy it is sobering to realize that the proof of its structure has had to wait until the twentieth century. Photography was the prior technical breakthrough required first to measure proper motions, second to establish the periods of the faint variable stars which yield the distance to the central bulge, and third to record spectra from which Doppler velocities can be ascertained. The final step, again dependent on photography, was recognition of the obscuring role played by interstellar smoke. Of course men have long speculated about the Milky Way, suggesting as much as two centuries ago that it might be an 'island Universe.' But the proof had to wait for a technical breakthrough. It gives one cause to wonder how much more of a fundamental kind remains undiscovered simply because our tools are inadequate.

If we understand the geography of the Galaxy, what then of the two Magellanic Clouds which so resemble fragments of the Milky Way that have broken off? If we focus our telescope on the large Magellanic Cloud, we find once again myriads of stars in the eye-piece. Study shows that these stars are identical to those in our Galaxy, but much fainter. For instance the Cepheid variables are 25 times fainter than those in our galactic bulge. Their colours are the same, so smoke cannot be responsible. We estimate, after using the inverse square law, that the Cloud must be $\sqrt{25}$ or 5 times as far away as the bulge. It follows that the Cloud is another enormous stellar system, another galaxy, about 150 thousand light years away. The Milky Way is not unique, though it is much larger than the Magellanic Clouds.

Two centuries ago William Herschel established that there are numerous nebulae in the sky, faint patches of light not dissimilar to the Magellanic Clouds, but much smaller and fainter. One such object, upon which we shall focus the telescope, is called the Sombrero Hat because of its shape. Seen in the eye-piece it is a most disappointing sight, a faint smudge of light with a bulge in the centre and a dark lane running across the middle. But is this not just what we would expect to see if we looked at the Milky Way edge-on from a long distance away? Unfortunately no individual stars can be seen in it, so we have no immediate way of establishing the distance. There is, however, another such object in the Northern Hemisphere, larger and brighter by far, called the Andromeda nebula (Plate 3). Andromeda looks to be the very twin of our own Galaxy, with a swirling disc containing dark bands of obscuration, and a brighter central bulge. But how big is it and how far away? As late as 1886 astronomers despaired of finding the answer. To quote from Sir Robert Ball, writing at the time:

> There is one problem of the greatest interest with regard to the nebulae, which astronomers often turn over in their thoughts, and which they as often despair of seeing satisfactorily solved. That problem is this—how to find the distance of a nebula. The difficulties of finding the distance of a star are so great, that it is only by a lavish expenditure of time and of patience that the work can be accomplished; but the difficulties are much greater and apparently insurmountable in the case of the nebulae, and not method has yet been devised which will enable us to solve this mighty problem. Our knowledge on the subject is merely of a negative character. We cannot tell how great the distance may be, but we are able in some cases to assign a minor limit to that distance. It is believed that some of these nebulae are sunk in space to such an appalling distance that light takes centuries before it reaches the earth. We see these nebulae, not as they are now, but as they were centuries ago. . . . We have reached a point where man's intellect begins to fail to yield him any more light, and where his imagination has succumbed in the endeavour to realize even the knowledge he has gained.

Ball's despair was to be dispelled thirty years later by the construction of a new generation of giant reflecting telescopes. Through these giants, individual stars, and in particular Cepheid variables, can be picked out in Andromeda, opening the possibility of finding its distance at last.

I do not want to anticipate a story that is told in a later chapter. Suffice it now to say that our Galaxy, far from being unique, is simply one among a myriad such objects which stretch out as far as the eye or telescope can see. And Sir Robert's surmise of intergalactic distances measured in light centuries was a gross underestimate. Andromeda, which turns out to be a near neighbour, is two million light years away. We must contend with a Universe immeasurably larger and richer than our great-grandfathers dared to conceive. The light from some of the galaxies we can see set out towards the Earth before there was anything more sentient here than microbes.

Progress in astronomy has been marked by the discovery of a series of ever larger structures, each one encompassing the largest structure that was known hitherto. Will this progression ever end? The tentative answer seems to be yes. So far as we can photograph there are as many faint galaxies in one direction of space as there are in any other, suggesting that the hierarchy of structures has finally come to an end. More persuasively, radio telescopes can now see back to an epoch before galaxies existed, to a time when the Universe was young and before its present-day structures had condensed out of the primeval broth. On that immense scale the cosmos appears to be smooth and homogeneous. We can speak of a youth and indeed of an origin for the Universe because, as we shall see later, Doppler measurements tell us that all the galaxies are flying apart as if the whole Universe were expanding. Once upon a time, many aeons ago, all we can presently see was crammed together into a superdense, superhot volume. There was a big bang, and ever since then it has been flying apart.

Such has been progress that the twentieth-century astronomer is faced with questions which his predecessors would have scarcely dared to ask, still less attempted to answer in any but theological terms. How old is the Universe? How large is it? How did the atoms in it come to form and what preceded them? From out of atoms how did structures like stars and galaxies condense, and why do they shine? What was the cosmos like during the first seconds of Time, and what will become of it in the end?

Our main preoccupation, the question we shall be concerned with in this book, is this: 'How much mass is there in the Universe, and of what does it chiefly consist?' In a sense it is the most fundamental question of all, because the answer to it contains the answers to many of the other questions outlined above. The mass density will control the amount of gravitation, which in turn will control the dynamics of the Universe. Given sufficient self-gravitation the present expansion will halt and the cosmos will one day collapse. The rate of expansion in the past determines how old the Universe is now. The present atomic composition, as we shall see later, is set both by the very early expansion rate and by the energy density present at the time. And following Einstein we now believe that the fundamental structure of space-time, that is to say the shape and size of the cosmos, is consequent upon its mass-content.

Our question is allied to, but notice it is not the same as, the simpler query: 'What can we see in the Universe?' We can see stars a long way off simply because they shine. We can see galaxies even further afield because they contain billions of luminous stars. But we have discovered that between the stars and galaxies lie stupendous voids of apparently empty space. Can we assume these regions are truly vacuous? Are we so fortunate that all the significant denizens of space have already called attention to themselves by emitting light, or some other detectable radiation? We know that during the day the Universe goes out; it is swamped by the day-time sky. How much is still swamped at night? After all, a modest amount of moonlight suffices to mask the Milky Way and indeed all the other galaxies. Even when the sun and moon have set we still find ourselves embedded in a giant Galaxy ablaze with stars rather like prisoners in a lighted cell. Should we expect to see anything more than the street lights when we look into the darkness outside? Finally we have to recognize that mass, that is to say the capacity to generate a gravitational field, is by no means confined to structures made of conventional atoms. Through the famous Einstein equations $E = mc^2$ (or $m = E/c^2$) raw energy alone has the property of mass. Heat has mass, light has mass, waves of any kind have mass, and there may be mass in overwhelming quantities locked up in weird particles like neutrinos which scarcely ever interact with the familiar fabric of matter.

The perceptible content of the Universe may thus turn out to be a minor constituent. Nevertheless it is a luminous braid that must be interwoven, through the force of gravity, with any invisible material. Neptune was found through the perturbation of the known planets. White dwarfs and neutron stars reveal themselves by swinging their visible companion-stars in space. And if the Milky Way is controlled by black holes or neutrinos then their gravitational presence will show up through the motions of its luminous stars. The observations are to hand. It is time to fathom what they mean.

Before we start, though, there are several matters of an almost philosophical kind which we should not ignore. We explore the cosmos by the application of laws established in earthly laboratories. Is it not presumptuous, even foolhardy, to assume that the laws of geometry and physics behave far out in the cosmos just as they do down here?

We have to start somewhere. We have to make hypotheses. But we can check on ourselves by following the hypotheses to their logical conclusions. We use them to make predictions and then check to see that the observations agree. An incorrect hypothesis will in the end usually lead to a paradox. For instance if, arguing from my hypothesis, I know that star A must be brighter than star B, when the reverse is observed to be true, then something must be wrong. There is an internal inconsistency, and that may be exciting. It tells us the Universe is not quite as we thought, and that is what astronomy is all about.

The continuous cycle of exploration, hypothesis, prediction and observational test is what distinguishes science from mere witchcraft. Scientific theories must be robust enough to stand up to a continuous battering from observations and experiment. And one can see now the importance of the quantitative and mathematical aspect of science. If I predict that the sun will rise in the morning you will not be impressed. But if I tell you it will rise at 06.17 precisely, and it does, then you might grant that I know a thing or two. The astronomer is continually checking his hypotheses against hard quantitative observations, and this should give some confidence in what he has to say, even when he is speaking of the remotest outposts of space-time.

Of course not all our hypotheses are consciously stated. Sometimes we assume a thing is so obviously true that it never occurs to us to check it. For instance men assumed without question for thousands of years that space was straight. Euclid even had a theorem purporting to prove it. And if space is straight it makes no sense to speak of a finite Universe, as we did in Chapter 1, when we called on a finite Universe as a possible way out of the Olbers' paradox.

A straight or flat space is one in which the familiar laws of geometry hold; one in which triangles have interior angles adding up to 180 degrees; one where parallel straight lines never meet. But if you draw triangles on the surface of a sphere the angles add up to more than 180° because the surface on which they are drawn is curved. Of course the Earth is such a big sphere that you can draw a rather big triangle on it before the difference from 180° becomes noticeable. So it could be with astronomical space, which is incomparably larger. If we draw out a really large triangle, say between three planets, one would need to carry out measurements before deciding what the sum of the interior angles really is. It turns out that space *is* slightly bent round the sun.

This opens the possibility that cosmic space as a whole is curved and finite. Start off in one direction and one could find oneself eventually coming back from the opposite pole, without encountering any wall or edge en route. We shall find that the curvature and finiteness of space is determined by the amount of mass within it, the very quantity we are anxious to find. Actually even if space is finite, containing a finite number of stars, it does not offer a resolution of the Olbers' paradox because given sufficient time, the stars within it will fill the cavity with radiation up to an intensity which will brighten the nighttime sky. The cosmos has to have a finite age.

The moral should be clear, even if it is all too easy to forget. No scientific hypothesis should ever be regarded as sacrosanct, no matter how respectable its pedigree. The notion of Ultimate Truth has a seductive appeal for which theologians and quacks may fall, but never a true scientist. His theories are ever on probation; they must continually pass through the testing crucible of experiment and observation. Indeed part of astronomy's attraction for the scientist is that the Universe offers testing conditions that can never be realized in the laboratory, extremes of temperatures, of density and gravity, where he can put his fondest theories on trial.

It is worth noting that astronomy is a 'softer' science than some because the astronomer cannot usually experiment with his subject matter. One cannot come back with a teaspoonful of a black hole and probe it minutely in the laboratory as the atomic physicist can. In partial compensation the astronomer has a vast and diverse playground in which to watch his theories succeed or fail. Nevertheless astronomical hypotheses must remain more provisional than most and astronomers must therefore expect more astounding surprises along the way.

If it turns out that the Universe is closed and finite, as well it might be, it may seem sensible to ask what could lie outside and beyond. The reply is that science can give no answers there, for by definition no information can reach us from Beyond. Such a Beyond is causally disconnected from our Universe and hence any speculations about it are entirely fruitless since they can never be subject to observational test. It does not matter anyway since, once again by definition, Beyond can never have any perceptible consequences for us.

Likewise it is natural to ponder what was going on before the big bang.

As we shall see later, we have quite respectable theories about what happened during the first few minutes of Time, and respectable physicists are now speculating about happenings in the first microtrillionths of a second. Eventually, however, one runs into a conceptual barrier at the point where conditions were so extreme that they can never be simulated in the grandest atom-smasher.

At that point the scientist reluctantly walks away, leaving the more brazen theologians to step in, because the scientist will not dabble in speculation he can never test out.

Despite their pretentions scientists are rarely as dispassionate as they would claim to be. After all, it would be inhuman to expect an individual to devote a life's work to studying a problem if he had no interest, no prejudices about the outcome. Real scientists, including the best among them, are very often emotional, imaginative creatures. And of all subjects, cosmology attracts the most fiercely partisan. Only beasts could remain indifferent to the origin, the structure and the fate of the cosmos in which they live. Only saints could resign themselves to never knowing the answers. The upshot has been that every civilization, almost every generation, has manufactured a cosmological story, dressed it in respectability and taught it to its children. Thus one tribe in India pictured the earth as a huge table, supported on the back of three giant elephants, which in turn stood on the shell of a gigantic tortoise. To the Egyptians the sky was a celestial Nile along which the sun-god Ra sailed from East to West. The Greeks had Zeus and his companions on Mount Olympus.

If the efforts of our predecessors now seem laughable it will not do to become complacent ourselves. If they did not know much it seems likely to me that some of the profound discoveries remain still for us to make. They will be made, if history is any guide, by those who question skeptically the contemporary picture. Remember that we know very little, as individuals, *at first hand*. Most of us do not know that the Earth is round. We have been told that it is. To emphasize the point, suppose you were transported back into the fourteenth century, wearing only pyjamas, and carrying no other possessions. Your appearance, to say nothing of your unfamiliar cleanliness, would attract immediate attention. A crowd would gather, and finding yourself the centre of curiosity you would naturally try to explain what had happened. After a fashion you would probably be understood, though your strange patois would only excite further amazement. At some stage 'authority' would come upon the scene, possibly in the person of a village priest or the sheriff, and you would be taken 'into custody.' Should you persist in your incredible story you would eventually be brought before an examining board. The challenge would be to prove that your story was true. If you failed it is likely you would be sentenced to suffer one of the various fourteenth-century fates reserved for madmen, heretics, witches or criminals. Armed with 600 years of superior knowledge you have the challenge of convincing a bench of sceptical, but reasonable men that you *have* come from the future. You have a week in which to do so. How could you succeed?

Try the challenge for yourself. You will find it very difficult, perhaps impossible. The fact seems to be that all the knowledge we have as individuals in incredibly secondhand, is of no use except in contemporary circumstances, and cannot be demonstrated. We rely absolutely on our sources. That being so, it is easier to forgive our ancestors their more extravagant beliefs and superstitions. Explorers of the cosmos are wisest who travel sceptically.

Our present cosmology owes its theoretical framework to Albert Einstein. Einstein explained gravity as a local disturbance or curvature in space-time caused by the influence of neighbouring material. The theory, which is supported by observation, not only admits of a curvature to space, it relates that curvature is an exact way to the palpable matter and energy present. The curvature is not arbitrary but is caused by the mass embedded in it. If we can measure the density of matter then the curvature is calculable. Conversely if the curvature can be observed we can infer the density of all material in the vicinity, be it invisible or not.

But gravity is a long range force so if Einstein is right then at every point in space there will be a net curvature which is the summed effect of all the gravitating bodies in the Universe. In other words space as a whole could be curved by its contents in such a way that the cosmos is finite.

Einstein's equations tell us that if the averaged out density of all the material in space exceeds a critical value of about 1 atom per cubic metre (10^{-29} gms per cc) then the Universe is finite or 'closed' as we say. If, on the other hand, it is less then the critical value, then space is still curved, but in a different way so that the Universe is infinite or 'open.'

Obviously we want to know how the actual density compares with the critical value. Intriguingly it turns out that the density of *visible material* is close to but below the critical value. Actually it is only two per cent as great but in astronomical terms, where we are used to dealing with factors of a trillion, then two per cent is very close, especially when you consider the huge uncertainties. As we have seen, astronomical distances are hard enough to come by and masses are if anything harder to find. In any case, as we have underlined many times, visible material may be only a part of the story.

Not only does the geometry of space hang on the critical density, but the dynamics and fate of its contents depend on the very same number. A dense Universe will slow its expansion and recollapse, while an empty Universe will go on expanding for ever. No wonder astronomers are fascinated by the density of space and intrigued by its closeness to the critical value. The difference between the two is called 'the missing mass' and many astronomers harbour the ambition to find it. Some of us find the notion of a closed Universe aesthetically appealing while others are as equally repelled by the notion that such a cosmos must inevitably collapse one day bringing all we know to an end. It is not surprising that more than intellectual passions are sometimes aroused, or that arguments rage far and wide. Our task will be to follow these arguments and find out where they lead.

While we have been talking, night in Chile has faded into dawn. The last stars have hidden themselves behind the veil of light. To the back of us the orange orb of sun peers over the Andean battlements, its beams painting the desert below with glinting fires. During the night sea-mist has crept up the valleys from the Pacific in long mucous tendrils which dissolve before our eyes. As the sun rises into his kingdom, all the Universe that was ours has fled. In place of the fanfare which surely the occasion demands we hear one tiny bird piping his defiance, or perhaps his wonder at it all.

As well as we can, from a mountain top in Chile, we have set the scene of our investigation. Our quest will be a search for no less than the missing ninety-eight per cent of the Universe. Our quarry is, if you like, the astronomer's El Dorado. The rumour of its existence depends on intricate theoretical arguments bolstered here and there by strong circumstantial evidence. The proof of its reality remains to be found. A feverish hunt is under way, and if it is to be uncovered at all, discovery could come any day now and from the most unexpected quarter. For unlike El Dorado the missing mass, if missing it is, exists not at the end of a dangerous voyage, but pervades and moulds the space which is all around us. The proof, if proof there is to be, could lie in the alchemy of the atoms, could be seen through the eye-piece of a telescope, could reveal itself through the symbols of an equation, could spew out of a computer, could in fact be staring us in the face. This book is the story of a cosmical treasure hunt. We shall hear rumours, follow up promising clues, keep alert for new ones. We shall weigh evidence and speculate on its application. We shall hear the protagonists in hot debate. But first of all all we must introduce the leading characters—the galaxies.

4

Galaxies Galore

SPACE of itself is invisible. If we would dissemble its properties and survey its dimensions we can do so only through studying the luminous objects which find their existence within it. And once we search beyond the confines of our own Milky Way system into the ocean of space beyond, nothing is to be seen save galaxies, galaxies more numerous even than the faintest stars, galaxies which stretch away in their myriad clouds and clusters to be lost as tiny dots against the night sky. On its largest scale the visible fabric of the Heavens is stitched in galaxies, and we cannot hope to understand very much without first unravelling their mysteries. But galaxies are complex entities with a rich variety of unexpected and bizarre properties. Despite fifty years of intensive observations we have to concede that very much of a fundamental nature remains unknown. Unlike the stars, which betrayed their simplicity, and soon yielded up their secrets to the theoretical physicist, the galaxies have kept close council with themselves. And we should do well to remember our ignorance at all times, for partial knowledge could easily lead us wildly astray. Nevertheless we know far more about galaxies than our grandfathers did, and if we do not know all of the answers, perhaps we have guessed most of the questions.

Apart from our own, only the two Magellanic galaxies far to the

South are easily visible to the naked eye. To see galaxies you need a telescope, but even then they appear merely as faint amorphous smudges of light. What luminosity they possess is spread over the sky in extended ghost-like images, images that can be seen only on the darkest of nights at the darkest of sites. The faintest hint of moonlight or man-made contamination, and they slip elusively away. This low surface brightness makes study of them very difficult and the reader should not be misled by some of the striking photographs of galaxies he will see in this and in other books. You can be sure these photographs have been selected and doctored, not dishonestly but to bring out some specific point. They disguise the real difficulty of extragalactic research.

From the earliest days of telescopic astronomy galaxies, or nebulae (clouds) as they were known to distinguish them from the point-like stars, were recognized and catalogued in many parts of the sky. The astronomers of old, who were obsessed by the search for comets wandering among the stars, regarded the fixed nebulae as something of a nuisance. And there were so many of them too. From their garden in eighteenth-century Windsor, William Herschel, his sister Caroline, and his son John discovered and described thousands. What were they?

Some resolved themselves in the telescope eye-piece into swarms or clusters of stars. Others, usually centred on a bright star or two, were, with the aid of the spectrograph, later shown to be composed of incandescent gas. But the third and most numerous class defied any such resolution or explanation. They were observed to shun the Milky Way and appeared in a variety of more or less regular shapes centred around their comparatively bright nuclei. For a century tireless eyes stared at the feeble diffuse lights, failing entirely to perceive any more revealing details.

The discovery of the full structure and variety of the extragalactic nebulae had to await the invention of photography and its adaptation, late in the last century, for use at the telescope. Though less acute than the eye, the photographic plate more than compensated by its ability to accumulate and store hours of light at a time. Moreover the final record could be measured as objective and indisputable.

The nebulae thus revealed were, if nothing else, at least unsurpassingly beautiful. They appeared to fall into three main classes. First there were the whirlpool or spiral nebulae, resembling nothing so much as gossamer catherine wheels in the sky. Their symmetrically structured dark and bright clouds spoke irresistibly of rotation (Plate 3). Next, and almost as common, were the so called elliptical nebulae.

More concentrated towards their nuclei, they were less flat (some were clearly spherical), sported no clouds either bright or dark, and appeared like giant star clusters (Plate 8). But no individual stars could be seen. Last and least common were the irregular nebulae, broken into bright and dark, like the spirals, but lacking their rotational symmetry.

Beautiful as these first photographs were, and rich in unsuspected and surprising detail, they entirely failed to settle the big question—what were these beautiful nebulae, and in particular, how far were they away? Indeed they fuelled a controversy which was to last for forty years. At its root the problem was quite simple. Nothing they contained was recognizably familiar. No comparisons could therefore be made and hence no distances could be inferred. Some astronomers claimed to see the spiral nebulae as embryonic planetary systems, commingling out of the pristine gas, about to become solar systems like our own. If they were right then the spiral nebulae were intrinsically very small and uncomfortably close. Others preferred a diametrically opposed interpretation wherein the nebulae were seen as quite distinct 'island Universes,' Milky Way systems in their own right, at immense distances from our own. To resolve the issue, better evidence was needed, and it was to be some time in coming.

The credit for solving the problem must go largely to George Ellery Hale. Hale was the visionary, the money raiser, and the telescope builder par excellence. The new photographic techniques cried out for bigger and better telescopes designed expressly to exploit them. Used to observing in the frozen cloudy skies of Wisconsin, Hale recognized the importance of clear mountain sites where the seeing was good. He raised money from wherever he could, largely from private philanthropists. To manufacture the largest reflecting telescopes ever built, he pushed the technology of his day to its limits. With the use of mules, the massive components were dragged up steep hillsides to be erected on Mount Wilson, in the coastal ranges of Southern California. Not only was this a site atmospherically superior to its predecessors, but it was sufficiently far south to observe much of the virgin and, as it turned out, crucial Southern Hemisphere. His 60-inch telescope saw its first light in 1907, to be followed in 1917 by the 100-inch, by far the most powerful instrument of its time. Before finally subsiding with nervous exhaustion, Hale searched about and recruited the brightest young men for his observatory: men like Harlow Shapley and Edwin Hubble. It was very largely Hale's boys, using Hale's telescopes, who were to crack the riddle of the nebulae. But, as so often in science, there were to be alarms and diversions upon the way.

Harlow Shapley was a self-made man of many parts. Almost on a whim he gave up his job as a small-town journalist to study astronomy. Shapley's manifold interests included the ants which can usually be found sharing any desert observatory. When Hale invited him to Mount Wilson, Shapley determined to find the distances to the remarkable globular clusters, of which about a hundred were known (see Fig. 16). These were not nebulae, for they clearly resolved into stars, of which they contain up to a million. Only in their directional distribution were they similar to galaxies, for both shunned the rich star-lanes of the Milky Way.

To get at their distances Shapley proposed to use the variable stars already known to inhabit these clusters. These variables, though fainter by far than variable stars like the Cepheids which lived in the Milky Way, were closely similar in other respects. The similarity suggested that the variables were all of the same ('Cepheid') type and hence intrinsic luminosity, but that the cluster variables appeared fainter only because the globular clusters were removed to a great distance. Such an assumption leads, via the inverse square law, to an estimate of the *relative* distance of the two populations of variables, those in the galaxy (galactic) and those in the clusters. But Shapley was interested in more than that. He wanted their *absolute* distance and for that purpose it was first necessary to gauge the distance and luminosity of the Cepheid variables in the Milky Way plane.

At the same time no one knew the Cepheid luminosities, and as they lay far beyond the range of parallax measurements Shapley had to devise a rough and ready statistical method for estimating their distances. The method assumes that the galactic Cepheids are in random motion relative to each other and to ourselves, as likely to be moving radially along the line of site as transversely in the plane of the sky. The radial velocities could be measured rather exactly with a spectrograph—using the blueshift or redshift. The transverse *angular* velocities become apparent as slight displacements when photographs taken years apart are compared. Now measured angular velocity is simply true transverse velocity divided by distance. If the true transverse velocities should be, as Shapley argued, equal *on the average* to the measured radial speeds, then the *mean* distances are determined at once.

Never mind about the details. The point is that the distance problem forced Shapley, like so many astronomers before and since, into some ingenious but roundabout and vulnerable arguments.

On reducing his painstaking observations, Shapley found that the galactic Cepheids were very distant and therefore highly luminous—ten thousand times as luminous as the sun. This in turn implied that the much fainter globular cluster Cepheids were a very very long way off indeed.

Armed with their distances and with their positions plotted on the sky, Shapley drew up the three-dimensional distribution of the globular clusters in space. If, as he believed correctly, they were associated with our Galaxy, moving about it in a flattened swarm in space, then our Galaxy was an enormous structure, with its centre 30,000 light years away in the bulge of the Milky Way near Sagittarius. The total diameter of 100,000 light years was far greater than anyone had dared to suppose hitherto, and the sun was tucked away far from its centre.

This radical re-assessment of scale caused a buzz of controversy which repercussed into the debate over the nebulae. If they were island Universes or galaxies with equal stature to our own, then they must reside at enormous distances, distances to be measured in millions, even tens of millions of light years. Even from the closest such nebulae the visible light had set out on its journey towards us long before the evolution of modern man.

These breathtaking distances set two very direct and apparently insuperable challenges to the Island Universe hypothesis. In 1917, Ritchey, another of Hale's protégés, spotted that a new star or nova had briefly flared up in one spiral nebula, until it assumed a brilliance almost equal to the parent nebula itself. Such novae as were known in our Galaxy were 10,000 times fainter than the Galaxy itself and so it seemed unlikely that Ritchey's nova could be so much brighter.

And then again, if the sizes of nebulae were to be measured in thousands of light years, there was no chance of comparing photographs taken a few years apart, and spotting their rotation. For such enormous structures to rotate perceptibly during the course of human history would imply speeds in excess of the speed of light, speeds considered unreached by matter. Yet Andreas van Maanan, also at Mount Wilson, had measured just such a rotation, and his measurements had been independently confirmed. Shapley concluded that the nebulae were either contained in our Galaxy or were close satellites to it.

But the Island Universers, led by Heber Curtis in a famous debate in 1920, were unrepentant. Nebulae were so obviously separate galaxies equal in every way to our own, they said, that Shapley's measurements were wrong, and the accuracy of his Cepheid method, and the assumptions on which it was based, were called severely into question.

With the opening of the great 100-inch telescope in 1917 unsurprisingly detailed photographs of nebulae accumulated at the Observatory. And on these photographs, over the next ten years or so, the crucial clues were to be found. Firstly, van Maanan's rotational measurements were discredited; not the last independently confirmed observations that astronomers have made and subsequently found to be wrong. Then Knut Lundmark, visiting Mount Wilson from Sweden, showed that novae come in two types, one ten thousand times more luminous than the other. With these objections removed, it was possible to reconcile the island Universe hypothesis with Shapley's distance scale. But definite proof was still entirely lacking.

Another visitor to the Observatory, J. C. Duncan from Wellesley College, made what was perhaps the germinal discovery. On photographs of the Andromeda nebula, which is much the biggest nebula in the sky, Duncan noticed what appeared to be very faint variable stars at the limits of possible measurement.

After Duncan had returned to his college to teach his students, Edwin Hubble, one of the staff astronomers with regular access to the telescope, took up the problem in earnest. Were those exceedingly faint dots really variable stars? If so, where they Cepheid variables like our own? For if they were, then the distances could be inferred using Shapley's Cepheid luminosity determinations.

In his search for variables over the next two years, Hubble took and examined literally hundreds of plates of a few bright nebulae like Andromeda. The dots were indeed variables which waxed and waned just like Shapley's Cepheids; variables whose distance could be deduced from their relative faintness. There could be no doubt about it. These, the largest and presumably the closest nebulae, were no less than a million light years away. They were enormous star-systems built to the same dimensions as Shapley's Milky Way. At last, and conclusively, we knew the answer. The nebulae were independent galaxies resembling our own.

Space and its denizens, revealed through tiny dots on a photographic plate, now outleapt the wildest spring of imagination. More, and more surprising, was soon to come. After Hale and his protégés, astronomy would never be the same.

We must follow up some of the main implications of the Mount Wilson discoveries. The new distance scale means that the galaxies are immense distances apart; between a hundred and a thousand galaxy diameters separate them. All but a billionth of the cosmic volume consists of apparently empty intergalactic void. I say apparently, for nothing substantial has been detected in it so far. No stars, no smoke, no gas, and only a little radiation. But in its vastness, thinly spread or secretly lurking, all of our missing mass could lie comfortably concealed. We shall return to look at what is known of this void in much greater detail later on. For now we must be thankful that at last it contains none of that obvious obscuring smoke which has so bedevilled the exploration of our own Galaxy.

Secondly, these immense distances face us at once with a fundamental difficulty. There will be no possibility of detecting any galactic motions transverse to the line of sight. For when translated into the angular motions we might perceive, transverse velocities of less than the speed of light will at these distances defeat all hope of measurement. On the cosmic scale the Universe will be, and is, in a busy dynamic state. But to our eyes, limited by the span of human existence, it is held in a perpetual frieze. Ours is a single still picture of the dance, and from that still we must try to reconstruct all that has taken place, and all that is to come.

Radially speaking, in the line of sight, we are not so handicapped. The blueshifts and redshifts of light perceived by our spectographs tell precisely of motions towards and away. There is too a welcome and surprising bonus. Light from the nearest galaxies takes a million years to reach us; hence we see them as they were a million years ago. The more distant Universe appears as it was billions of years in the past, a significant interlude in the span of cosmic history. By comparing far with near we are seeing past against present, with the clear possibility of deciphering cosmic evolution.

From the apparent brightnesses of galaxies, the distances allow us to infer their true or intrinsic luminosities. For a biggish galaxy like our own, the total radiant output matches that of ten thousand million suns. And if it is largely composed of stars like the sun, the number it contains must be ten billion likewise. But we should qualify this figure in two ways. To begin with, galaxies come in a wide range of size and luminosity, from monsters more luminous than our own, to dwarfs scarcely one ten-thousandth as bright. Secondly, the true luminosity of a galaxy is all but impossible to measure, for galaxies have no sharp edges and merge imperceptibly into the haze of the background sky. One might suppose that the outermost parts contain very little more light, but that is not so. In many cases, as the measurements are pushed outward to greater and greater radii, and to lower and lower levels of surface brightness, there is no clear sign of convergence. As the surface brightness falls away so the corresponding emitting area increases, and the convolution of the two stubbornly continues to increase. Of course this divergence, of which we shall have more to say later on, cannot continue indefinitely. If it did, an infinite luminosity would be implied. For the moment we simply note that the measured luminosities are merely lower limits, and not as we would wish, upper bounds.

Of what are galaxies made? The nearest galaxies, like the dwarf Magellanic Clouds and the giant spiral in Andromeda, are sufficiently close to be studied in some detail and compared with our Milky Way. Most of their light is emitted by a population of stars apparently not so very different from our own. Of course only the very brightest stars can be identified at these distances, while the main sequence stars which emit the majority of light and certainly contain the vast preponderance of mass are too faint to be seen. All we can do for certain is compare the most spectacular stars in extragalactic nebulae with the most spectacular stars in our vicinity. And indeed we do see the blue and red supergiants, the luminous Cepheid variables, the novae and so on, indicating that stellar physics out there follows very much the same course as it does around us.

Besides individual luminous stars, we can also identify larger structures like globular clusters, associations of hot, young stars and incandescent gas, and clouds of obscuring smoke which are familiar closer to home.

While the bulk of the stars, main sequence dwarfs, like the sun, are invisible as individuals, this silent majority speaks of its presence in the integrated or bulk light from the parent galaxy. For instance, the overall galactic colours indicate a familiar mixture of blue, yellow and red stars, both giants and dwarfs. The composite or averaged spectrum of a galaxy is very informative. It is rich in spectral features right through from the blue to the red and can tell us much about the luminosity, temperature, relative numbers, age and chemical composition of the contributing population. A single such spectrum contains thousands of pieces of information which can be disentangled with enough painstaking care. Within a spiral galaxy, we see evidence of a stellar mix which is closely comparable to our own. If we are right in our supposition that stars are basically simple creatures, controlled in their destiny by only two of the most certain laws of physics, then this comparability should come as no surprise.

The gaseous content of galaxies can be likewise discerned using a number of separate techniques. Incandescent gas heated by very hot stars in the neighbourhood show up as bright features in the visible spectrum. Cool, absorbing gas in the line of sight produces characteristic dark features. The bulk of this gas is, as it is in our Galaxy, hydrogen, and hydrogen emits a sharp feature in the radio spectrum which radio telescopes can map as to position, velocity and temperature.

In some galaxies, particularly spirals, obscuring smoke shows up as obvious dark lanes. In others, where lanes are less remarkable, its presence is betrayed by obvious reddening of the intermixed stellar light. But although the gaseous and smoky component of galaxies is rather noticeable, it hardly ever comprises more than 20% of a galaxy by weight, and in general the fraction is nearer 5%.

The gaseous fractions, the colours, the spectra and the stellar populations which give rise to them, vary considerably from galaxy to galaxy. Interestingly, though, these gross properties correlate very well with the immediate appearance, or to use the technical jargon, the morphological type of a galaxy. The elliptical galaxies are uniformly red, they contain no young blue stars and virtually no gas or smoke. They would appear to be populated entirely by very old red stars with the majority of light coming from red giants. Unfortunately we have no giant elliptical galaxies in our neighbourhood, which leaves a number of uncertainties as to their detailed stellar content. Although ellipticals are otherwise uniform, they come in a very wide range of ellipticals with the power of 10^{11} suns, while the faintest dwarfs are likewise ellipticals, but ten thousand times fainter.

At the far extreme from the distinctively smooth and regular ellipticals, we find the so-called irregular galaxies like the large and small Magellanic Clouds. As their name would suggest, they are patchy and irregular in aspect. This is mainly explained by their comparatively large content of gas, up to twenty per cent by mass. Where the gas has recently condensed into stars, superficial eruptions of young blue supergiants spot the adolescent faces of these galaxies, disguising a more regular and much older population underneath. The young blue stars dominate the integrated spectra, making irregulars the bluest galaxies of all.

The beauty queens of the galactic chorus line are unquestionably the spiral nebulae. Their discus-like shapes, decorated with spiral necklaces of blue stars, excite as much admiration as wonder. When we take our eyes off the jewelry we find a complex anatomy underneath. The central bulge is the nucleus of a wider distribution of old red stars which stretches out spherically to the full diameter of the system. Spinning within this halo is the more obvious disc of main sequence stars which contain most of the weight and which, despite the photographs, provide most of the light. The photographs, generally taken in blue light, selectively highlight the young blue stars. These are condensing out of the gas which, in small amounts, forms spiral traceries within the massive disc. These disc galaxies differ one from another in two chief and clearly apparent ways: in the ratio of bulge to disc, and in the relative admixture of gas. In the rather common lens galaxies, the gas and young stars are almost totally absent, whereas in spirals like our own they can make up ten per cent or so by mass. Very small dwarf spirals are quite unknown and in luminous output most are giants like the Milky Way.

Of the galaxies we can easily see, about forty per cent are ellipticals, forty per cent are discs and spirals, while the remainder are irregulars.

As we earlier explained, any motions of galaxies transverse to the line of sight, though they certainly exist, will not be measurable. But the in-and-out components of velocity readily show up as shifts in the spectral features, both at optical and radio wavelengths. From these shifts we can build up a picture of the motions both within galaxies and as between one galaxy and its neighbours.

The speeds concerned are typically measured in hundreds of kilometres per second. These may sound high but in terms of the distance scales we are dealing with they are actually stately. Spinning at such a rate a galaxy requires no less than a quarter of a billion years to complete a single rotation, while in direct translation a galaxy typically moves only one galactic diameter in 100 million years, or 100 galactic diameters in the total lifetime of the Universe. Since nearest neighbours are generally a hundred diameters or so apart, isolated galaxies must remain at home in their own territory.

The spectra of disc galaxies confirm our suspicion that they spin about their centres. More unexpectedly, the irregulars are mostly rotating too. The elliptical galaxies, however, are more of a puzzle. They do rotate, but at much slower speeds than their flattening would seem to indicate. What we chiefly see is evidence that the stars within them are buzzing about randomly like bees in an angry swarm. All the nebulae are kept from breaking apart by self-gravitational forces, or so we believe.

The conformity between a galaxy's outward appearance and its internal composition is both striking and puzzling. Ellipticals never contain much gas, but why not? Irregulars are bluer than spirals, which in turn are bluer than ellipticals, but why? One feels there are underlying physical relations between the three sorts of galaxy, but what are they? Could it be that they represent succeeding ages of a single developing population, the irregulars being young, the spirals older, and the ellipticals oldest of all? Apparently not, for ancient stars are seen in all three types. Is there a single underlying property, the amount of rotational angular momentum for instance, which controls both outward appearance and inward constitution? None has been found so far. Or have they formed by quite separate processes, the spirals from gas, the ellipticals from pre-existing stars? For the moment we cannot say.

The student of galaxies finds the subject bristling with hard and interesting problems. To put our own investigation into perspective, that is to give a feeling for the gross uncertainties that abound in the background, I shall outline two problems.

Problem one. How did galaxies form? Observations do not help us for, as I mentioned above, embryonic galaxies are not found in our neighbourhood. Bereft of direct evidence, we must resort to theoretical speculation. But in 1944 a Russian called Lifschitz proved mathematically that galaxies cannot condense out of an expanding Universe. No one since has found a convincing way round Lifschitz.

Problem two. How do galaxies relate to their nuclei? In the very cores of about one per cent of galaxies we find active nuclei, that is to say nuclei which radiate X-rays, radio waves, infrared light and so on. These nuclei are no more than a light day across. We can be sure of their size because the radiation they emit varies on a timescale of a day, and nothing can change significantly in less than the light travel-time across it. At that size nuclei occupy only one part in a thousand billion billionth of the whole galactic volume. But in rare cases, called quasars, the tiny nuclei can radiate no less than a thousand times more energy than all the 10^{11} surrounding galactic stars put together. Are we to understand that it is the nucleus which dominates the galaxy, and that galaxies as we have described them are merely fur upon the beast? But what then of the commonality of galaxies like our own whose nuclei, if they radiate at all, contribute very little to the total luminosity? We have to ignore these uncertainties in the hope they will not affect our investigation.

What about the sociology of stars and galaxies? Within our Galaxy stars show a strong tendency to cluster. As often as not stars are observed in binary pairs, in triplets and in higher multiplets right up to the million stars or more found in globular clusters. Such aggregations are not the result of chance or transitory associations; they are far too common for that. They are the visible manifestation of physical structures held together for long periods by the force of gravitation. Indeed we believe that stars cannot form in isolation but are born in large clusters, a thousand or more at a time. But over the course of ages the more weakly bound clusters break up either via slow evaporation or through the tidal effects of more massive structures like the galactic bulge. An isolated star like the sun almost certainly belonged to such a cluster in the distant past.

If stars are clustered, what about galaxies? If we plot out the positions of the brightest thousand or so galaxies on the celestial sphere, we see at once that they are not uniformly spread (Fig. 3). On the contrary.

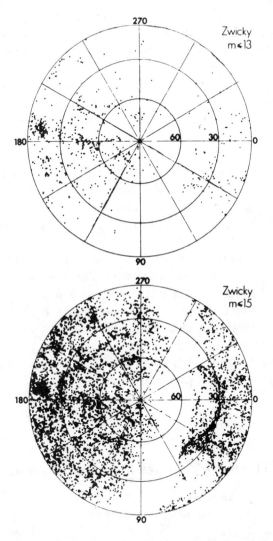

Fig. 3 The distribution of the brightest galaxies projected on the North Celestial Hemisphere, with only the very brightest galaxies in the top figure, and including those 6 times fainter in the lower. Notice that the galaxies are not smoothly distributed, but are strongly clustered with the centre of the giant Virgo supercluster very noticeable at extreme left. The blank zone is the region obscured by our own Milky Way.

While some regions of sky remain virtually barren, other regions are endowed with a rich constellation of bright galaxies in close associations. While some associations are no more than the chance superposition of background and foreground nebulae, the clustering tendency is far too strong to be anything but a real physical association. Further investigation confirms that immediate impression. The radial velocities of the members show they are moving through space at the same speed, and at the same distance from us, held together by mutual gravitation. The clustering comes in all ranges of size, from binary pairs of galaxies orbiting each other in close attendance, through small groups with perhaps a dozen members, or up through associations of a hundred galaxies, right up to the richest clusters and superclusters with ten thousand members or more. Plate 6 shows a rich cluster in the constellation of Columba. To my eye it is awe inspiring to see hundreds of galaxies, each one of the same scale as our entire Milky Way, strewn about through space like so many pebbles on a beach. The precise number of galaxies is unknown, but you can count over a thousand and there may be many more that are either too faint to be seen, or too small to be distinguished from foreground stars.

So strong is the clustering tendency that it is difficult to find truly isolated galaxies. For instance our own Milky Way is one of two dominant spirals in a group of about twenty galaxies, about two million light years in extent, the group itself belonging to a far larger supercluster whose centre is situated fifty million light years away in the constellation of Virgo. To give some idea of the typical galaxies to be found, and the range in their properties, Table 4.1 gives some details of the known members of our local group.

TABLE 4.1

The main galaxies in our local group

Object	Distance thousand light years	Luminosity in suns	Remarks
Andromeda	2,200	2×10^{10}	Giant spiral
Our Galaxy	—	1×10^{10}	Giant spiral
Messier 33	2,300	3×10^{9}	Less luminous spiral

TABLE 4.1

The main galaxies in our local group (*cont.*)

Object	Distance thousand light years	Luminosity in suns	Remarks
Large Magellanic Cloud	160	2×10^9	Dwarf irregular companion to our Galaxy
IC 10	3,600	8×10^8	Isolated irregular
Small Magellanic Cloud	180	4×10^8	Dwarf irregular companion to our Galaxy
Messier 32	2,200	3×10^8	Dwarf elliptical companion of Andromeda
NGC 205	2,200	3×10^8	Dwarf elliptical companion of Andromeda
NGC 6822	1,600	1×10^8	Dwarf irregular
NGC 185	1,900	9×10^7	Dwarf elliptical companion of Andromeda
NCG 147	1,900	7×10^7	Dwarf elliptical companion of Andromeda
IC 1613	2,200	6×10^7	Dwarf irregular
WLM object	2,200	6×10^7	Dwarf irregular
Fornax	800	2×10^7	Dwarf spheroidal galaxy
Leo A	3,600	2×10^7	Dwarf irregular
Sculpter	270	4×10^6	Dwarf spheroidal
Draco	330	2×10^5	Extreme dwarf spheroidal

N.B. There are in addition another dozen extremely faint galaxies believed to lie in the group, most of them extreme dwarf spheroidals like Draco or dwarf irregulars like Leo A.

To outward appearances the vast majority of visible mass in the cosmos is locked up in galaxies. The 'missing mass,' which is our prey, represents the difference between this visible density and the larger hypothetical density required to close the Universe. Our first priority is therefore to estimate the visible mass, averaged over a representative value of space, of the galaxies we can see. And this problem subdivides into two stages, which will be tackled in detail in some of the following chapters.

First we must weigh individual galaxies, according to their type and luminosity. How much mass is there in a giant elliptical, how much in a dwarf irregular, and so on? Answers can be found using a number of techniques, none of them easy or totally satisfactory. For instance we can weigh individual nearby stars of different luminosity and colour. If in addition we can estimate from the total luminosity and colours of galaxies how many stars of each type they contain, then we find an estimate of the total galactic mass. Then again, there are dynamical techniques. To hold themselves together against the centrifugal forces caused by internal rotation, galaxies must exert strong binding forces through the agency of self-gravitation. But gravitation implies mass, so the rotational speeds in galaxies lead to estimates of their masses. Alternatively, since galaxies are so often found in pairs, in groups, and in clusters, we can seek to measure, via their relative separations and velocities, the gravitational influence of galaxies upon one another, and this yields mass-estimates too.

Once we have built up a reasonably satisfactory picture of the way the masses of galaxies relate to their luminosity and type, we can proceed on to stage two. The problem here is to choose a volume of space representative of the Universe as a whole, and to estimate the numbers of galaxies of all the different types and intrinsic luminosities to be found in it. This is relatively easy for bright galaxies because they are clearly seen out to a great distance. It is much harder for dwarf galaxies which are so difficult to observe that even today we are still finding new ones in our own immediate neighbourhood. It is our old enemy observational selection once again.

Supposing that allowance can be made for all these awkward effects, we can complete our calculation. We select our representative volume of space, count up all the individual galaxies within it by type and luminosity, multiply by the separately estimated masses, total it all up and then divide by the volume to compute our density: the mean density of visible material. We shall find a number which is below, but tantalizingly close to, the magic figure required to close the Universe.

But from where does that magic, hypothetical density come? In the next three chapters we aim to find out.

Gravitation

As we have seen, the Universe is a very 'grainy' place. Instead of being evenly spread, the visible material is gathered into relatively dense lumps, called galaxies, separated from each other by enormous volumes of apparently empty space. Indeed the visible galaxies occupy only one billionth of the total volume. On a smaller scale the galaxies themselves are gritty too, being composed of stars, each with the average density of water, separated once again by vast interstellar distances. Only one part in 10,000,000,000,000,000,000,000 of the galactic volumes is inhabited by stellar material. Why such an overwhelmingly grainy structure?

The answer, in short, is gravity. Each piece of material attracts all the neighbouring material towards itself and what might once have been a relatively homogeneous soup is now largely condensed into dense massive lumps or grains. The Earth is squeezed into its spherical shape by gravity. Gravity holds the planets in orbit about the sun. Because gravity compresses the core of the sun to such high temperatures, it radiates. The stars of the Milky Way are constrained to move in circles about the galactic centre by gravity. And wherever we see galaxies aggregated into clusters, we hold gravity responsible once again. If then we are going to learn anything about the structure of the Universe on the scale of clusters or smaller, we must understand a thing or two about gravity.

Furthermore, the force of gravity presumably does not stop when you reach the boundary of a cluster. Clusters must be attracted towards one another and indeed the whole cosmos must exert, and experience, the compressive force of self-gravitation, acting upon itself. This will lead us, via a dilemma, to some important cosmological considerations. In fact gravity provides one of the main arguments for suspecting a missing mass.

And if there truly is such a mass of invisible material, our best hope of detecting it lies through the gravitational effects we might observe on those pieces of matter which we *can* see. Altogether, gravitation is going to feature so prominently in our whole debate that we must step back and trace man's conception of it.

To understand some phenomenon in Nature, the experimentalist tries his best to isolate that phenomenon from all extraneous circumstances. By disentangling it from all supernumerary complications, he hopes to see that phenomenon in its purest form, and so to perform repeatable experiments with repeatable results. Once these results are known, various hypotheses can be advanced to account for them, until one can be found which both accurately describes the existing experiments and suggests new ones. A strong theory is one which accurately predicts very much more that is new and unexpected. A bad theory predicts incorrectly. A weak theory describes little more than is contained in the original experimental material upon which it was based, predicting very little or nothing new.

Our knowledge of the laws of motion was long in gestation, precisely because it proved impossible to isolate a moving test-body from all manner of extraneous influences, particularly gravity. Galileo, probably the greatest experimentalist of all time, carried out the crucial experiments by rolling heavy cannon-balls about on a wooden plane. Of course, gravity is not removed, but because the upwards force of the flat plane exactly balanced the downward gravitational force on the balls, then gravitation could be largely discounted. Galileo discovered that a ball, if set in motion, and then left alone, would continue to roll along the plane in a straight line at a constant speed. Not a profound discovery, you might suppose. But it was revolutionary for it showed, contrary to the assertions of the ancients, that steady motion in a straight line required no external influence to keep it going. Causative agents were only to be looked for if the rolling ball deviated or accelerated from its steady rectilinear progress.

Newton erected a daring generalization upon Galileo's simple cannon-ball experiments. He asserted that any body, removed sufficiently far

from outside forces, will continue indefinitely upon a steady rectilinear path. It was daring because no one had ever seen such a completely
isolated body. Indeed, no one has done so even today. The nearest we
can approach such an abstraction is for an astronaut to carry out experiments with floating cannon-balls in free-space. But even there
some weak gravitational forces remain. What Newton saw was that the
further removed were the unavoidable external agencies, the more
nearly did the observed motion approach his idealized steady form. He
therefore took the logical step and argued that in the hypothetical limit
where all influencing material was removed to a greater and greater
distance, the motion of a body would approach more and more closely
a steady rectilinear form.

On the face of it most of the observations were against him. In the
laboratory, unsupported bodies accelerate violently downward, while
in space freely moving bodies like the moon and the planets execute
elliptical motions around a parent body. But Newton countered these
objections by supposing that a long-range force existed between massive bodies operating in free space, a force which he called gravity. He
was able to show, in a series of wonderful mathematical investigations,
that provided the hypothetical force increased in proportion to the
masses of the bodies concerned, and decreased in proportion to the
square of the distance between them, then all the observations could
be very accurately explained. But he went much further than that: he
asserted that gravitation was a universal phenomenon, not confined to
bodies of astronomical size, and that any two pieces of material of mass
m_1 and m_2, separated by a distance d_{12}, would experience a mutual
force of attraction F_{12}, such that:

$$F_{12} = G \, \frac{m_1 m_2}{d^2_{12}}$$

where G is a numerical constant which can be measured once and for
all, and which describes the strength of gravitation. The Universal
nature of this supposed force led Newton to make a number of new
and dramatic predictions, all of which were to prove triumphantly
successful. He calculated the relative masses of the sun and planets for
the first time, explained the tides at last, accounted for the bulge at the
Earth's equator, and predicted the return of comets. Though not without its critics, the theory appeared irresistible, and on its shoulders the
good Sir Isaac rose to be first among Science's Gods.

Newton's first law of motion states that the natural motion of a free
body is steady movement in a straight line. (It makes no sense to say

that the natural state is motionless, since motion, or lack of motion, is a purely relative concept.) If then, in the behaviour of a particular body, accelerations or deviations are detected, it is to be supposed that forces of some kind are at work. His second law defines those forces in terms of the mass m and acceleration $\underline{a}$ of the body involved. (Deviations, being but transverse accelerations, are subsumed into the notion of acceleration.) He then asserted that the force F is given by:

$$\underline{F} = m \times \underline{a}$$

($\underline{F}$ and $\underline{a}$ are underlined to emphasize that they are, unlike mass, *directional* concepts.) It is natural that the mass should enter, for everyday experience (try push-starting a car) informs us that the heavier a body, the more force required to get it going.

We are shortly going to criticize Newton's theory, but before doing so we should acknowledge just how effective it has proved to be. Indeed it is astonishing to realize just how much that was previously a mystery has been explained and put to use by such a simple idea. The fall of an apple, the acceleration of the moon, the orbits of the planets, the mass and density of the Earth and its companions in space, the ebb and flow of the tides, the discovery of Neptune, the appearance and reappearance of comets and eclipses, the maximum height of mountains, the set of the oceanic streams, the development of mapping and navigation, the launching of satellites, and the design of space-colonies, all fall into Newton's simple scheme. A single theory, first guessed at by a young man of 22, has changed the world out of all recognition. In doing so it has relegated sorcery and superstition to the shades of history, and promoted in their place the scientific approach, which has proven to be more profitable as well as provocative. No event in history, not the birth of a nation or the decline of a civilization, no battle of revolution, has exceeded in its impact Newton's theory of universal gravitation.

Notice that the theory does not 'explain' gravitation, it merely describes its mathematical form. Newton did not attempt to explain his universal force in terms of a more primitive first cause. He simply assumed its existence and offered an unambiguous recipe for calculating the magnitude and direction of the force in any conceivable situation. Thereafter he was content to let the theory stand or fall on its observable consequences. And so it stood, very successfully, for 250 testing years.

From the very outset, however, there were critics and objections. Indeed many of Newton's contemporaries and successors found his ideas very hard to swallow. They were being asked to believe in a virtually intangible force which acts across vast distances of space, with no discernible intervening agency. Moreover, the force was presumed so weak that it could not be (in those days at least) demonstrated as acting between normal bodies in a laboratory experiment. And if such force acted directly on the straight line between Earth and sun, why was it not cut off during an eclipse when the moon intervened? Furthermore, there is something distinctly fishy about it. As Newton's second law of motion has shown, the mass of a body represents its inertia, or resistance, to acceleration. If two masses are subject to the same force then the more massive will react more sluggishly. But in the Earth's gravitational field, all bodies, irrespective of mass, are found to accelerate at the *same* rate. (Once the supernumerary effect of air resistance is removed by carrying out experiments in a vacuum, heavy and light bodies, cannon-balls and feathers, are found to fall to the ground at identical rates.) Newton, they said, had 'fixed' this difficulty by arguing that the more massive a body the stronger the gravitational force acting on it. While that may be legitimate, and seems to work exactly in practice, they felt that a good philosophical justification was needed for his argument. After all, why should the gravitational response of a body have anything to do with its mass, which is a quite separate concept and a measure of its resistance to acceleration? The fact that it all seems to work in practice is not, they said, a satisfactory reply to these objections.

While that may be so, most practical scientists were not too impressed by these largely philosophical considerations. They saw their task as the largely empirical one of making correct predictions. Judged in that light they found Newton's theory to be a great success. Only if the opposition could come forward with an alternative theory of equal predictive power would they lend much credence to such subsidiary and more philosophical arguments. In 1915, Einstein was to propose just such an alternative theory.

How Newton's ideas came to be superseded by Einstein's around 1915 is a long and technical story, for which we have neither the space nor the necessity in this book. But it is worth examining one or two key points of difference between the ideas of the two men.

Newton's laws are implicitly embedded in his notions of straight or Euclidean space, and of an absolute time dimension which can be agreed upon by all observers. And neither notion stood up to certain mathematical and scientific developments which took place in the nineteenth century.

To begin with, it was recognized that one cannot mathematically prove, as Euclid had asserted, that space is everywhere 'straight' or 'flat.' Whether it is or not can be determined only by careful observation or measurement. Although we are not properly equipped to perceive or imagine curved space directly, space could still be curved and we can define clear experimental ways of measuring that curvature. For instance, a triangle in space is made up of three intersecting straight lines, where a straight line is defined to be the shortest distance between two points. Take three points in space and stretch strings between them, so defining a triangle. Now *measure* the internal angles of the triangle very carefully (see Fig. 4). If they sum to exactly 180°, or two right angles, then the space in which the triangle is embedded is 'flat' or 'straight,' because the geometry inside it behaves like ordinary Euclidean geometry carried out on a flat surface. If they sum to more than 180°, we say the space is positively curved, and the geometry behaves like that on the surface of a sphere. If they sum to less, we speak of the curvature as negative. Whether we believe our own space is curved or not must be determined by experiment. In our neighbourhood the curvature must be exceedingly small, because it cannot be

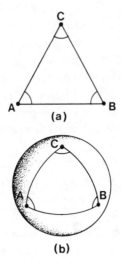

Fig. 4 The geometry of triangles in flat and curved space. (a) Triangle ABC is drawn on a flat surface and one can show by measurement or argument that the three interior angles add up to 180° or two right angles. (b) Triangle ABC is drawn on a sphere. The lines AB, AC and CB are 'straight' in the sense that they are, on the surface, the shortest distances between A and B, etc. We see, however, that the interior angles add up to more than 180°.

measured directly. Even so, it may suffice to explain gravitation (see later). There is a close analogy to the curvature of the Earth's surface which, while it is not readily apparent in everyday life, nevertheless exists, and can be determined by triangulations made over sufficient distance. If indeed space is curved, then the distance squared appearing in Newton's law is no longer defined so unambiguously and will depend on the curvature.

Secondly, Einstein was influenced by the theory of electromagnetism, a theory which developed very successfully in the nineteenth century thanks largely to Michael Faraday and James Clerk Maxwell. Moving electrical charges and magnetic poles influence one another by inverse square forces very similar to the force of gravitation. But the electromagnetic theory is not based on action at a distance as was Newton's. Space is thought of as being threaded throughout with electrical and magnetic tensions. Even when no movements or accelerations can be seen, the stresses remain, and their size and direction can be mapped out, in the case of magnetism for instance, by dropping iron filings onto the surface surrounding a bar magnet. Such a theory is called a field theory, for it confers properties upon the field of space surrounding electrified bodies, the bodies interacting with one another only through the agency of the field. If gravitation were to be similarly cast as a field theory, one would not need to ask why it is that the attraction between sun and Earth does not vanish when the moon is interposed during an eclipse. This is because gravitation would then be a property of the general field surrounding all those bodies; the interposition of the moon would not be significant, at least not so dramatically as if the force were thought to act in a straight line between the sun and ourselves. For such reasons, Einstein was looking for a field theory of gravitation.

Another nail was to be hammered in Newton's coffin by some modifications in our conceptions of space and time. Einstein realized that such modifications were forced upon us by developments in electromagnetism. To apply Newton's law it is first necessary to specify the exact position of two bodies, and the distance between them, at each precise moment of time, and so long as space and time are absolute that can be done. But if the measurement of space and time intervals depends somewhat on the motion of the observer, as Einstein averred, then Newton's law could no longer be applied unambiguously.

By 1910 these developments had made the construction of a new theory of gravitation mandatory, and Einstein set out to produce one. He was looking for a theory which (a) could be used unambiguously by all observers irrespective of their state of motion; (b) was a field theory like electromagnetism; (c) differed in its observable consequences very little from Newton's which, for all its theoretical weaknesses, had proved to be so accurate in practice; and (d) was as elegantly simple as possible.

The origins of Einstein's theory are to be found in two conceptual weaknesses in the earlier hypothesis. The first of these is the notion of force acting at a distance with no perceptible intervening agency. The second is the idea of force at all when what is observed is that all bodies, irrespective of their mass, in the same gravitational field *accelerate* at the same rate. Newton, you remember, avoided this difficulty in a legitimate but artificial way by supposing that the gravitational force on a more massive body increased in exact proportion to its inertia or mass. In other words, he had to assume that the two conceptually quite distinct properties of matter, its gravitational effect and its inertial effect, were exactly proportional to one another.

Einstein started from an opposite point of view. Like Newton he began with the observation that in regions far removed from other matter the natural tendency of a body is to move through space in a straight line. If then, in the space near a large mass like the sun, a body is seen to move in a curve, then he supposes that in such regions space itself must be curved. A planet is still trying to move straight through space, but space, being curved, causes it to deviate its path into an ellipse; and since it is the *space* which is affected, it is quite natural that all bodies embedded in it at the same point should be curved or accelerated in the same way. Thus there is nothing surprising about a cannon-ball and a musket-ball falling at the same rate.

If gravitation is to be attributed to curvature, and if gravitation is only observed to act in the vicinity of massive bodies, it follows that the curvature of space must be caused by the action of the heavy bodies embedded within it. What is needed is a mathematical relation between the amount and type of curvature, and the amount and distribution of matter in the neighbourhood. And on the basis of the four precepts enumerated as (a) to (d) above, Einstein was able to construct such a relation which he called his 'field equation.' On the left hand side of the equation is a complex measure of the curvature, denoted by C_{ij}, and on the right an equally complex expression T_{ij} which incorporates all the needed information on the local distribution of mass. This equation is then written:

$$C_{ij} = - KT_{ij}$$

where K is some undetermined constant. To determine K, Einstein pointed out that as Newton's theory is so nearly right in all but the 'strongest' gravitational fields, his own (Einstein's) equation must reduce exactly to Newton's in such cases of weak gravitational fields as, for example, exist near the Earth. From this argument he found that K $= 8\pi G/c^2$, where G is Newton's gravity constant, and c is the speed of light (300,000 km/sec). Einstein's theory can then be stated in two simple laws:

(1) A body will move through space in as straight a line as it can;
(2) Space is curved by the presence of nearby matter and the amount of curvature C_{ij} is given by:

$$C_{ij} = -\frac{8\pi G}{c^2} T_{ij}$$

Before examining the implications of Einstein's new theory we should raise three technical issues which will return to haunt us later on.

The first thing to be said is that Einstein's law is mathematically far more complicated than Newton's, and this difficulty is compounded by our inability to imagine curved space intuitively. For this reason practical scientists stick to using Newton's inverse square law whenever the required accuracy of their calculations will allow. That is to say they will stick to the Newtonian law whenever the gravitational fields are 'weak.' A 'weak' field is defined as one for which the pure number

$$\eta \text{ (pronounced 'eeta')} = \frac{2GM}{c^2R}$$

is much less than one; M being the mass of the heaviest body around, and R being the distance from its centre. η is a measure of the escape velocity from the gravitational field in question, compared to the speed of light. It is in fact the square of this ratio. To escape from the sun's field near the Earth calls for a velocity of 300 km/sec or one thousandth that of light. Hence η near the Earth is one thousandth squared, or one millionth. In the sun's field near the Earth, $\eta = 2$ parts in one hundred million or 2×10^{-8}. For calculations where an accuracy of no more than one part in 10^7 is needed, that is to say for all practical purposes, Newton's theory is quite adequate. It is used, for example, in all spacecraft calculations. Even at the surface of the sun $\eta = 3 \times 10^{-6}$, so

Newton's theory is usually sufficient there too. But for white dwarf stars $\eta = 10^{-3}$, for neutron stars $\eta = 10^{-1}$, and for black holes $\eta = 1$. Even if we discard Newton's law on grounds of principle, we shall continue, for reasons of simplicity, to use it in practice, except when we are dealing with the Universe as a whole, because then η rises to be of significant size.

Secondly, we must reiterate that in its mathematical form Einstein's law is closely and deliberately modelled upon Newton's, and indeed at great distances from the gravitating body of interest, the two laws become identical in their practical predictions. If then we have reasons to doubt the accuracy of the inverse square law at such great distance, Einstein's law must fall under suspicion too.

Thirdly, it must be said that the mathematical construction of Einstein's field equation is not unique. Other, but more complex, curvature equations have been devised, but there is no experimental evidence to support them as yet, we shall ignore them henceforth.

In the end, though, however brilliantly conceived, however distinguished their originators, theories are all the fallible constructions of fallible human minds. The reliance we can place upon any of them must ultimately rest on experimental tests and observations. Empirical evidence is the only objective judge as between Newton and Einstein. So what does the evidence tell us?

Since, in the weak gravitational fields with which we are mostly familiar, Einstein's theory is deliberately constructed so as to agree with Newton's, finding experimental evidence to distinguish between the two cannot be simple. Even in the field close to the sun η is only of the order of one part in a million, which argues that distinguishing measurements will be correspondingly small and therefore difficult to make. Nevertheless half a dozen distinguishing tests have been proposed, and carried out, with results which strongly favour Einstein.

The first, and perhaps most direct, is concerned with the deflection of light by the sun. Light, we know, travels in straight lines. But if, in the vicinity of a massive body like the sun, space is slightly curved as Einstein would have us believe, then light rays passing through it will be seen as slightly deflected. As the sun makes its annual progression across the background stars, the starlight should accordingly be slightly bent so that the observed angular separations of stars lying close to the line of sight of the sun will vary slightly as the sun passes by. In practice the necessary observations are difficult to carry out, both because the predicted deflections are very small and because stars can be photographed close to the sun only during the rare and brief moments of a total eclipse. Since light of energy E possesses a mass of E/c^2, Newton's theory likewise predicts a deflection, but by an amount only half as great. A clear test was finally devised in the mid 1970s using the relative deflections of radio stars which can be observed without the necessity for an eclipse. The results argued strongly in favour of Einstein.

The planet closest to the sun, and therefore the one subject to those stronger gravitational fields where the two theories most diverge, is Mercury. Even before Einstein its orbit was found to deviate ever so slightly from the perfect Newtonian ellipse. Each orbit, whick takes 88 days to complete, fails to close up with the last by about 1 part in 2 million, i.e. the ellipse appears to rotate very slowly about 1 millionth of a degree per orbit, and Einstein's theory, but not Newton's, accounts for this exactly.

Other, less convincing tests are known. Of recent years hyper-dense white dwarf and neutron stars, with truly strong gravitational fields (η between 10^{-2} and 10^{-1}), have been studied carefully with results which broadly favour Einstein. It is now generally conceded that as between the two theories of gravitation, Einstein's is much the more likely to be right.

Why should such minutiae concern us, preoccupied as we are with the large-scale structure of the Universe? To begin with, η, as measured for the whole visible volume of the Universe, is not small. Suppose we can see out to a distance d, then as the volume of a sphere of radius d is $^4\pi d^3$, the mass enclosed in that visible sphere is the volume multiplied by the density ρ. So:

$$\eta = \frac{2GM}{c^2R} = \frac{2GM}{c^2d} = \frac{2G}{c^2d}(\frac{4}{3}\pi d^3\rho) = \frac{8\pi Gd^2\rho}{3c^2}$$

Since G and c are known, we have only to estimate d and ρ. If we take for d the most distant objects we can see, and for ρ the density of visible galaxies averaged out over the whole explorable volume, we find η to be .01 or about 1%. If the invisible density is high, η will be correspondingly higher. And if we can increase d significantly, as we hope to with the launch of a telescope into space in 1986, η will be of order 1 or 100 per cent. In such circumstances it will be of prime importance to understand which theory of gravitation actually applies, for the two theories give rise to quite divergent predictions.

The second motive for preferring a theory of gravitation which, like Einstein's, incorporates the notion of a space with non-Euclidean properties is that it allows us to conceive of a finite Universe, but one with no boundaries. Without curvature one is forced to the view either that the Universe is infinite or that it possesses some definite edge or wall. And as Giordano Bruno pointed out long ago, the idea of an edge is ludicrous.

But the most satisfying aspect of Einstein's theory, from a cosmological viewpoint, is that it provides, through the field equations, a link between the properties of space and the material content enclosed therein. Space is no longer an arbitrary God-given construct which can assume any one of a number of bizarre forms. Its properties stem from, and are defined by, its visible and invisible contents. In principle this elevates cosmology from the purely theological province up into the domain of observational science. If the contents can be ascertained, the geometry is defined. Conversely, if the geometry can be measured, then we can ascertain the true material density.

Even so, by itself Einstein's theory does not solve all our fundamental problems. For instance, the Olbers' paradox remains. Even if the Universe is finite it will, given sufficient time, fill up with radiation from its constituent stars until the sky is very bright. It is true that the time required is very long, 10^{24} years in fact, but it is finite nonetheless.

Let us briefly summarize what we have learned about gravitation. Gravitation manifests itself as an acceleration of neighbouring bodies towards one another in space, and results in the grainy structure observed in the Universe. Newton described that acceleration as a mutual force of attraction between any two masses, which increases in proportion to their masses and decreases with the inverse square of the distance between them. On the other hand, Einstein saw the acceleration as the effect of a curvature imposed on the local space by the embedded masses. Although very different in conception, Einstein's law is deliberately constructed so as to agree with Newton's in the limit of a weak gravitational field, where strength is defined by the number η on a scale from 0 to 1. Defined thus, most familiar gravitational fields are very weak indeed and it matters not which of the two theories we use in practice. Careful measurements show that where the two theories differ slightly in their predictions, Einstein's is the more accurate. However, Newton's law is mathematically much simpler, so we generally prefer it in most astronomical applications. But on the scale of the observable Universe as a whole, η differs significantly from zero, so that the curvature of space is no longer a local irregularity but a property of the Universe in its entirety, related to and defined by its overall material content. But if curvature removes the difficulties inherent in imagining a boundary to the Universe, it resolves neither the Olbers' paradox nor the propensity of any self-gravitating body like the cosmos to collapse upon itself.

6

The Expansion of the Universe

WE have seen that gravity suggests a collapsing Universe while the Olbers paradox precludes one of infinite age. Both arguments were reasonably well established by 1870, so it is somewhat surprising that no one dared to take them to their logical conclusion or conclusions. This failure of nerve would be all the more unexpected if there were anything to those accusations of over-presumption which are sometimes levelled at cosmologists by laymen. In fact the accusations are quite misplaced, for as Weinberg points out: '. . . our mistake is not that we take our theories too seriously, but that we do not take them seriously enough. It is always hard to realize that these numbers and equations we play with at our desks have something to do with the real world. Even worse, there often seems to be a general agreement that certain phenomena are just not fit subjects for respectable theoretical and experimental effort.' Because of this timidity, cosmologists failed for sixty years to look for the missing piece in the cosmic jigsaw puzzle. When, in the late 1920s, the expansion of the Universe was forced upon their attention by observations, the discovery came not as a source of relief but as a matter for great surprise.

The late nineteenth century saw the introduction of two new observational techniques which were to revolutionize astronomy. The first of these was the replacement of the human eye-ball by the photo-

graphic plate, the second was the development of astronomical spectroscopy. These, together, led to the vital breakthrough.

A direct photograph pictures the light from a whole field of stars or
galaxies at once. In spectroscopy, however, the stars or galaxies are
studied one at a time, and the light is split up into its constituent
columns with a prism before being focussed onto the photographic
plate. A spectrogram therefore contains far more information about the
star or galaxy we are observing than a straight photographic plate.
Instead of a single image the spectrogram now contains a large number
of images side by side, each formed in a different coloured light.

We have mentioned already that because of their low surface brightnesses galaxies are difficult enough to observe even when all their light
is integrated. So to image them in hundreds of separate colours at once,
which is what spectroscopy amounts to, is very hard indeed and several nights of exposures were often needed with the early spectrographs to accumulate even a single exposure. Nevertheless interest in
galaxies was such that Vesto Slipher began a systematic survey of
spectra of the brightest galaxies in 1912 at Lowell Observatory in the
dark, clear skies of Northern Arizona.

When examined the spectra were found to contain dark lines similar
to those seen in spectra of stars like the sun. For instance, down in the
purple-blue region two prominent dark lines can be seen which are
known to correspond in wavelength to certain transitions in the calcium atom at 3933 and 3968 nanometre wavelength. In the case of the
sun, these lines occur because calcium atoms in the cool solar upper
atmosphere absorb out light at this wavelength from the hot radiating
surface underneath. Stars with different temperatures exhibit radically
different patterns of lines.

Galaxy spectra are found to contain far fewer features than their
stellar counterparts, but what features there are can be identified with
those in the commoner stars nearby to us. It is inferred that galaxies
contain a multitude of stars of different sorts, so that very often the
dark lines from one type of star are cancelled out by the light from
others, and that only a few features that are prominent in a wide variety
of stars can be easily discerned in the light from the galaxy as a whole.

One also notices that the lines of galaxies are much broader, i.e. they
cover a wider interval of the spectrum than in individual stars. This
broadening we attribute to the individual motions of the multitude of
stars within the galaxy acting through the Doppler effect. That is to
say, the spectral features of stars moving towards us in the line of sight
are shifted slightly towards the blue, while light from stars moving

away from us is shifted slightly towards the red. The summed effect of many stellar blueshifts and redshifts is to cause a broadening of each recognizable feature in the galaxy spectrum about its central wavelength.

Now Slipher noticed that in addition to broadening in his galaxy lines there was often a systematic shift of all the lines towards either the red or blue, and these shifts were attributed to systematic motion of the emitting galaxy as a whole. Typically these shifts implied motions either of approach or recession amounting to a hundred or more kilometres per second. Thus the Andromeda galaxy was found to be moving in our direction at 300 km/sec. These motions are hardly surprising when we realize that relative speeds of the same order are to be found among the stars of a single galaxy such as our own.

The surprise came as more and more galaxy spectra were painstakingly gathered. More spectra were found to be redshifted than blueshifted. Moreover, blueshifts of more than 300 km/sec were never seen, whereas Slipher registered some redshifts as high as 2000 km/sec. On the face of it, it appeared that the majority of galaxies were moving away from our galaxy. What was going on?

The alarm bells were slow to ring, if only because the early data were so fragmentary. Nevertheless, had Slipher's evidence been sooner juxtaposed alongside the cosmological arguments that were then under way, the expansion of the Universe might have been recognized rather earlier than it was. The problem then, as it is today, is that very few people have the brains and energy both to understand esoteric mathematical arguments and to learn all that is involved and implied in observational astronomy.

Following the construction of Hale's giant 100-inch telescope in California in 1917, one of his assistants, Milton Humason, was able to start accumulating galaxy spectra more readily, and these could be combined with Hubble's simultaneous measurements of galactic distances at the same Mount Wilson observatory. There is some argument as to who spotted the great discovery first: Lundmark or Hubble. But it does not really matter, for as the distances and redshifts accumulated there was no avoiding a momentous conclusion. The more distant a galaxy appeared to be, the more rapidly it receded from us. The fainter a galaxy of a given intrinsic luminosity appeared, the greater was its redshift. And the relation between recession speed and distance is as simple as possible; it is linear. That is to say the speed is twice as great at twice this distance. Superimposed on this general recession are the individual random velocities of galaxies which muddy up the story a little. For the very near galaxies that Slipher was working on, the recession effect scarcely amounts in many cases to more than their general random motions, so it is not surprising he did not spot the expansion first. Working with a much larger telescope, Humason and Hubble could observe galaxies out to distances where the general recession swamped the random motions completely. Plate 7 shows a convincing montage of their observations.

The law of recession of the galaxies, Hubble's law as it is called, can be stated simply and succinctly. If we ignore the small random velocities possessed by individual galaxies, any galaxy in the Universe will recede away from us with a velocity V directly proportional to its distance away d. In fact we can write

$$V = H_o d$$

where H_o is a number, called Hubble's constant, which can be obtained from observations.

At first sight it would appear that our Galaxy is selected out for special treatment and that all other galaxies are trying to avoid us like the plague. However, a little reflection shows this not to be the case. Consider four galaxies labelled O, A, B and C, lying at equal intervals along a straight line. Judged by an astronomer on A, O and B will be receding at velocity W, while C, being twice as distant, will, according to Hubble's law, be receding from him at 2W. According to him he is at the unique centre of a general expansion. But an astronomer situated on galaxy B would come to exactly the same conclusion. He thinks of himself as being at rest while A and C recede at speed W, while O being twice as distant recedes at 2W. In other words, the linear form of Hubble's law singles out no particular galaxy. All galaxies recede from one another like the fragments of a single cosmic explosion hurling its debris out into space. From whatever fragment or galaxy you look, it would appear that all the others are receding uniformly away.

So the whole visible Universe is expanding and this meets our earlier objection to the impossibility of a static Universe in the presence of gravitational fields. Even so it is a momentous conclusion to arrive at considering it was reached merely on the basis of microscopic displacements or redshifts in a few dozen extragalactic spectra. Is the Universe really expanding, or could not the spectral shifts be attributed to some other and unsuspected cause?

To answer this question it is best to follow through some of the implications of the expansion hypothesis. If the galaxies are truly receding now, then in the past they must have been much closer together than they are today. Indeed, if the expansion has continued uniformly then at some finite time in the past they must all have been crammed together in a single lump of very high density. All the perceptible structures in the Universe must be of an age less than some finite age T_0 when all structure would be obliterated by the super-dense condition near the outset. How old is the Universe judged in this sense, and how old are its constituent structures by comparison? If any are found to be older than T_0, the expansion hypothesis will have to be abandoned.

T_0 is easy to estimate from Hubble's law. If a galaxy distant d from us has been receding at speed v since the very beginning, than at a time T_0 = d/v ago it would have been crammed on top of us. But by Hubble's law d/v = constant for all galaxies = $1/H_0$. So the age of the Universe, under the expansion hypothesis, is simply

$$T_0 = 1/H_0.$$

From their observations Hubble and Humason found that galaxies recede at a speed of 160 km/sec for each million light years of distance away from us. Now a million light years at 160 km/sec takes 2×10^9 years or about two thousand million years, and so this was their estimate for the age of the Universe. Since radioactive dating of the oldest rocks proved the Earth to be at least 4 thousand million years old, there was a serious contradiction and much consternation. But as no other totally satisfactory explanation for the redshifts has been found, either then or since, astronomers adhered to the expansion hypothesis but with some uneasiness as to the accuracy of their observations. The uneasiness mounted as detailed calculation of stellar evolution showed that some stars are even older than the sun and hence older than the Earth.

To measure H_0, and hence T_0, it is necessary to obtain the recession speeds v and distance d of a sample of receding galaxies. Then, inverting Hubble's law, we find immediately that H_0 = v/d. Provided one is willing to accept the Doppler explanation for the redshifts, then the velocities can be found by direct measurement of the spectra. The distances of galaxies are, however, very much harder to find. As we pointed out in Chapter 4, they can be reached only after a long process of delicate measurements, careful comparisons and intricate arguments. From the parallaxes of close-by stars, we calibrate the colours and luminosities of stars on the main sequence. By comparing the luminosities of a few Cepheid variables with main sequence stars of the same velocity, Shapley then determined the absolute luminosity of the Cepheid variables as a function of their pulsation period. By locating Cepheids in the immediately neighbouring galaxies and measuring their periods and apparent luminosities, Hubble could find distances in a few cases by applying Shapley's Cepheid calibration. Unfortunately, Cepheids can be located only in galaxies that are so near to us that the recession effect hardly stands out above the random galactic velocities. To estimate distances to galaxies whose spectra exhibit unmistakable signs of the general recession, Hubble was forced into even cruder comparison techniques. For instance, he compared the apparent luminosities of the few very brightest stars in each galaxy, assuming, or rather hoping, that he was comparing stars of exactly the same intrinsic luminosity in each case. In this arduous and extremely delicate task, where one is always working at the limits of measurement, and where assumptions have to be made which may seem reasonable but cannot be fully checked, it is no surprise to find that mistakes can and have been made, and that Hubble's original value for T_0, the age of the Universe, was ten times too small. Even fifty years, and a lot of telescope time later, we cannot be certain of T_0 to better than a factor of two.

The largest upward revision of T_0 came about 1950, when T_0 was increased by a factor of ten. Work done with Hale's ultimate masterpiece, the giant 200-inch telescope completed on Mount Palomar in 1952, and work done elsewhere in locating Cepheids in star clusters where their distances could be more accurately assessed, resulted in an improved value for T_0 of around ten billion years. We know now that the earlier figure was largely in error because Shapley had not allowed sufficiently for the observation of smoke in the line of sight to his Cepheids within our Galaxy. The mistake was more than excusable in Shapley's time, for many of his successors made exactly the same error in confirming his results.

Our best estimates for H_0 today yield values in the range of 30 (± 15) km/sec per million light years, where the bracketed value gives some idea of the expected uncertainty. Following the launch of the Space Telescope in 1986, we are hoping to define H_0 much more exactly. T_0 now falls in the range 15 ± 5 billion years, which means there are no longer strong grounds for suspecting that the expansion hypothesis conflicts with the known ages of any cosmic constituents. However, it is worth emphasizing that ages are very difficult to come by in practice. The rocks and meteorites of the solar system we can date by radioactive means. For the ages of the stars we have to compare the observations with complex calculations of stellar evolution, calculations which in turn depend on some assumptions about their initial chemical composition. There is some confidence in the calculated ages of youthful or middle-aged stars like the sun, but our confidence progressively diminishes as we work backward to the oldest known stars which we think are to be found in globular clusters (10^{10} years). As to the ages of galaxies, we have very few clues, though it is usually assumed that they must be older than any of their constituent stars, where the hypothesis is that stars have formed only after the galaxies were born a considerable time ago.

Taken literally, the present expansion implies that about 10 billion years ago all the matter of the Universe was squeezed into a very dense mass, when the physical conditions must have been quite different from those of the present day. Squeeze matter very hard and it heats up. Conditions must have been far too hot and too dense for any stars or galaxies to have been present. Indeed, if we go back far enough one can imagine temperatures too great even for atomic nuclei to have survived. We have to think of a super-dense, super-hot soup lacking in any large-scale structures and composed of elementary particles like electrons, protons and neutrons transforming back and forth into one another in an intense bath of radiation. If that were truly the situation then surely some evidence is left today of the super-dense state.

With one notable exception, theoreticians failed once again to take these implications seriously. The exception was George Gamov, an imaginative Russian-American physicist who, among his many accomplishments, wrote the Mr. Tompkins series of popular science books. In 1949 Gamov published a paper analysing the early birth-pangs of the Universe, suggesting how, and in what proportion, the elements must have condensed out of the primeval soup, and predicting that relic radiation left over from the initial big bang should now permeate the Universe in all positions and in all directions. Unfortunately, some details of his analysis were found to be unsound, and so Gamov's pioneering paper was consigned with many others to gather dust unnoticed on the shelf. I say unfortunately because in its broad conclusion the paper appears to have been right, and Gamov's relic radiation could have been searched for and found much earlier than in fact it was.

It was to be fifteen years before Dicke and Peebles of Princeton University were to take the matter of the super-hot early Universe seriously once again. Unaware of Gamov's early work, they were led once again to the idea of relic radiation left over from the big bang. They calculated its likely characteristics and started building a special wireless receiver to detect it, for it turned out that most of the relic radiation should appear in the microwave radio spectrum.

But they were to be pipped at the post, and quite by accident, by two radio engineers at the Bell Laboratories in Holmdale, New Jersey, named Arno Penzias and Robert Wilson. Penzias and Wilson were using an antenna specially built for satellite communications to measure the intensity of radio waves emitted by the hot gas in our own Galaxy at heights well above the Milky Way. To their surprise they discovered an additional source of radiation which was isotropic, that is to say it came equally from all directions. At first they suspected a fault in their circuits but when, after careful testing, no defects could be found, they wrote a paper modestly entitled 'A Measurement of Excess Antenna Temperature at 4080 Megacycles.' When the two groups came to hear of each other's work through the grape-vine, they collaborated, and two papers were to appear side by side. The first, from the Bell Laboratories, described the observations; the second, from Princeton, gave the explanation for it in terms of relic radiation left after the big bang, as the earlier phase of the Universe was soon to be called. Much subsequent work has confirmed both the measurements and the explanation, and in due course Penzias and Wilson received the Nobel Prize for their momentous discovery.

What was it that Penzias and Wilson had detected? If we take the expansion quite literally, then we can wind back the clock and use the known laws of physics to calculate conditions as they must have been at each preceding epoch of the past. Because of the finite velocity of light, when we look outwards we are looking backwards in time to the physical conditions as they existed then. Thus we can check calculations against observation.

How far out can we look, how far backward in time can we see? It is comparatively straightforward to calculate the temperature of both radiation and matter as a function of density and time in the past. Using the Bell Laboratories result, we can then show that when the Universe was about a billion times as dense as it appears to be now, that is to say when the overall density equalled the density within present day galaxies, the temperature must have been about 3000°C. At such temperatures the hydrogen atoms which seem to make up the bulk of all cosmic material would have been dissociated into their constituent protons and electrons. Now free electrons are strong absorbers of radiation, so that at 3000°C and above the Universe must have been opaque. We cannot expect to see beyond that temperature domain. But for all succeeding epochs, when the free electrons were combined within hydrogen atoms as they are today, the Universe could become highly transparent. If today we look outward and back, we expect to see the 3000° stage of the Universe, the stage when it was last opaque. We expect to be surrounded on all sides by a hot 3000° blanket of gas.

That we are not roasted alive in such a furnace temperature comes about through the redshift effect. The hot, opaque blanket that we can see is situated in a region of the Universe that is now so far away that it is receding away from us at a speed very close to that of light. Each photon or wave reaching us from there is Doppler shifted or redshifted by a factor of no less than a thousand. Since stretching a wave by a thousand is equivalent to weakening or cooling it by the same factor, the opaque blanket seems to us to radiate not at a temperature of 3000°, but at a mere 3° above absolute zero. The radiation has been shifted right out of the visible spectrum where it was first radiated down into the microwave region, where only radio receivers can pick it up.

Incidentally, matter at 3000° does not radiate at a single frequency or wavelength; radiation is emitted throughout the visible spectrum from the ultraviolet into the infrared with an apportionment between the various wavelength regions specified by the laws of physics. It follows that the redshifted radiation which we observe must be correspondingly distributed throughout the microwave region, and recent observations taken by a high-flying U-2 aircraft confirm that this indeed is the case. Alternative explanations for such an isotropic microwave background are hard to think of, so there now seems little doubt that the Universe really is expanding out of a super-dense, super-hot initial state. Penzias and Wilson's accidental observation has proved to be the cosmological Rosetta Stone.

Although opacity stops us from peering directly backwards beyond the 3000° curtain, there is nothing to prevent us calculating what conditions must have been like at far earlier epochs. Because complex structures like stars and galaxies could never have survived in the hotter and earlier phases, everything was controlled, we believe, by a few simple physical laws. Astrophysicists have dared to calculate backwards to within one hundredth of a second of the moment of cosmic conception. And out of these calculations have come two remarkable predictions, which have an important bearing on our story.

Under the intensely hot conditions which must have prevailed in the early Universe, radiation and matter would not have behaved as they do under everyday circumstances. Such were the energies involved that nuclei of normal elements were broken into their most elementary constituents such as protons, neutrons, electrons, positrons and neutrinos, while the radiation was mostly in the form of destructive gamma rays, and the bizarre reactions which take place in such super-energetic conditions can be studied in modern-day particle accelerations. For instance, we would get the following common transmutations:

electron + positron   gamma rays
proton + electron   neutron + neutrino

where the double arrow signifies that a reaction can proceed in either direction. Not all imaginable transmutations can take place, for experiment has shown that certain conservation laws must be obeyed. For instance the total number of baryons (protons and neutrons) has to remain fixed both within a single reaction and overall. Likewise the number of photons (particles of radiation) and the number of leptons (electrons, positrons and neutrinos) are both fixed for all time. But within these constraints considerable variation of composition can take place, and the balance between one type of constituent and another is determined at any instant by the prevailing density and temperature.

The present-day Universe appears to be made up, by weight, of about 75% hydrogen, 25% helium, with all other elements totalling less than 1%. The Earth, with its heavy concentration of silicon, oxygen and iron and so on, is so unrepresentative, we believe, because lighter elements like hydrogen and helium boiled off in its early youth. Can the expansion scenario explain the present chemical composition of the Universe? If we are going to take the aforementioned calculations seriously, then surely it should.

The main ingredients of such a calculation are pretty straightforward. Besides the laws of particle physics, as determined by experiment, we must feed in the relative number of baryons and photons present. Since the conservation laws fix these numbers for all time, they can be estimated from the present-day matter-density (baryons) and the background radiation temperature (photons) using the Penzias-Wilson result. Lastly we feed in the observed God-given expansion rate, modified slightly to take account of the slowing effects of gravitation. Nothing else is necessary. The current chemical composition is irrelevant because nuclei were all broken into elementary particles in the hot early phases. Even the finiteness or otherwise of the Universe is of no account. This follows because there had been insufficient time for knowledge of the outermost parts of the Universe to affect any local nuclear reactions.

Granted these simplifications, let us follow through the results of the calculation, concentrating our attention almost extensively on the relative numbers of protons and neutrons present at any time, for it is these baryons which make up most of the mass of the Universe, and it is their ratio which affixes the chemical composition. The main reactions are:

6   antineutrino + proton   positron + neutron

neutrino + neutron   electron + proton

$\left.\begin{array}{}\\\\\end{array}\right\}$ (A)

neutron   proton + electron + neutrino   (B)

where the double arrows mean the reactions can proceed in either direction. Reaction (B) is very slow, for an undisturbed neutron takes on average about 10 minutes to decay, whereas in the dense hot conditions of the early Universe reaction (A) happened billions of times/particle/second. For (A) it is important to realize that the neutron is marginally heavier (by one tenth per cent) than the proton. Since $E = mc^2$, slightly more energy is required to form neutrons from protons than vice versa. So long as the primal soup was hot, i.e. energetic enough, this did not matter and reaction (A) proceeded in both directions with equal facility. However, once the temperature fell below ten billion centigrade, neutrons decayed more rapidly than they could be replaced and protons began to predominate.

If we connote the mean distance between baryons by d(t), then as the Universe expands, d(t) increases as a function of the age t. The density of matter obviously falls as $1/[d(t)]^3$ and it can be shown that the temperature, which dominates the various reaction rates, falls even more rapidly as $1/[d(t)]^4$.

This is what happens, we believe:

(a)   Between the zero and $t = 1$ second the temperature exceeds ten billion and the reactions (A) proceed equally in both directions yielding equal numbers of protons and neutrons. Reaction (B) is far too slow to be significant. No heavier nuclei can survive.

(b)   Between $t = 1$ second and age 200 seconds the temperature, and hence the available energy per particle, has fallen to the point where the more energetic neutrons are not being exactly replaced. At the end of this phase neutrons make up only 13% of all baryons, the remaining 87% are protons. A few heavier nuclei are beginning to form by reactions like these:

proton + neutron deuterium nucleus + photon (C).
then proton + deuterium nucleus helium 3 nucleus   (D).
and   helium 3 + neutron stable helium 4 nucleus   (E).

where the rates depend entirely on the production of deuterium in the first place by (C). However, (C) is reversible and at temperatures above 900 million degrees the deuterium is evanescent so all the rates are very low.

(c)   At t= 200 seconds the temperature falls below the critical 900 million level and suddenly reaction (C) proceeds all one way. The free neutrons, now forming only 13% of all baryons, are suddenly gobbled up into deuterium which, by the further reactions (D) and (E), is mostly transformed into stable helium. The few remaining free neutrons decay by process (B).

(d)   Between t = 200 seconds and t = 700,000 years, the temperature is too low for any further nuclear reactions, and at t = 700,000 years it has fallen to 3000°, when the electrons either combine with protons to form hydrogen atoms, or with the helium nuclei to form helium atoms. There being no free electrons left over, the Universe, which was earlier opaque, now clears and can be penetrated by observational techniques.

Two inferences are quite clear. The final proportion of helium is precisely determined by the fraction (13%) of free neutrons left to react, once the temperature has dropped sufficiently (to 9×10^8 degrees) for deuterium production to start. Since there are 2 neutrons and 2 protons in each helium nucleus, helium should remain to form about $2 \times 13\% = 26\%$ of the Universe by mass. This figure, depending as it does only upon particle properties, should be highly insensitive to the large-scale properties of the Universe. Before this prediction, which was made independently by Peebles and by Waggoner, Fowler and Hoyle, astronomers had already established that the observed helium composition was surprisingly uniform wherever it could be measured, and close to 25% by mass. This is a remarkable confirmation not only of the general expansion hypothesis but of our detailed understanding of its early and otherwise unobservable effects. No more remarkable demonstration of the power of human thought could be imagined.

Besides hydrogen and helium, only trace amounts of deuterium and helium 3 are produced in the big bang. Evolution is too rapid to cook up any heavier atoms like oxygen and carbon. Our second inference must therefore be that these 'heavies' are formed from the pristine hydrogen and helium by quite other processes later on. We now believe they are slowly cooked up in the cores of evolving stars, producing as a by-product most of the observed luminous stellar energy. When, at the end of their evolution, some of the more massive stars blow up as supernovae, trace amounts of these heavier elements are dispersed into the interstellar gas out of which subsequent generations of stars and planets will form.

Since the early thermonuclear cooking processes were largely independent of the large-scale properties of the Universe, we can hardly expect the present chemical composition to tell us much of the overall cosmic structure. But there is one notable exception. The amount of free deuterium left over depends critically on the density of baryons and since, as we shall see in the next chapter, it is the baryon density which probably determines the finiteness or otherwise of the Universe, a good clear deuterium measurement may be of great significance.

A direct demonstration of expansion would be an observation which showed that galaxies, and related objects, were crammed together more closely in the past than they are now. By counting galaxies per unit area of sky as a function of distance, or redshift, we should see the expansion directly at work. Simple as the idea is in principle, it is difficult to carry out in practice. At distances where the crowding should be unmistakable, galaxies become too faint to observe with our current optical techniques. They are lost as faint dots of light below the foreground brightness of the sky. Nevertheless, sophisticated indirect methods have developed which give information along the same lines.

Some few galaxies, radio-galaxies we call them, are powerful emitters of radio radiation. Because radio waves are of long wavelength (of order metres), relatively coarse engineering techniques allow us to build enormous receiving antennae up to 1000 feet in diameter. Such antennae, or radio telescopes, can detect radio galaxies at far greater distances than their much smaller optical counterparts. Unfortunately, however, the absorption and emission features seen in visible spectra are absent in the radio, so radio observations are powerless to measure the redshifts or distances to radio galaxies directly. To get round this, radio astronomers simply count the number of detected radio sources per unit of sky, as a function of the perceived radio luminosity. If indeed radio galaxies were more closely crowded in the past then the more distant, that is to say the fainter radio sources, should be over-represented by comparison with those nearby. Much effort has been invested in simple radio counts like this, mistakes have been made, and there have been several heated controversies. But there can be no argument today that the counts of radio sources as a function of strength are not commensurate with a static, unevolving Euclidean Universe. There are far too many faint sources to fit the simple picture, and if nothing else we can be certain that radio sources were crammed together more closely in the past.

More direct tests along the same lines can be undertaken with quasars or QSOs ('Quasi-stellar-objects') as we prefer to call them nowadays. These faint stellar objects, first unearthed in the 1960s, are found to have optical spectra with a few very strong emission lines in them. Because of these easily recognized lines redshifts can be measured for much fainter QSOs than for galaxies with ordinary spectra. In this manner QSO redshifts as high as 3.9 have been observed. These imply that QSOs are pathological galaxies undergoing violent outbursts in their nuclei, outbursts which increase their total output by as much as a thousand, outbursts which allow us to observe these strange beasts out to an enormous distance. Counts of QSOs as a function of their redshifts, and then presumably of their distances, indicate once again that QSOs were almost certainly crowded together far more in the past than they are now.

With expansion so satisfactorily confirmed, we have for the first time a model of the Universe which meets all the objections raised against earlier conceptions. Let us see how the model overcomes the hurdles one by one.

In the presence of gravity a static Universe was found to be impossible. An initially static Universe of the presently observed density would collapse into a super-dense lump in a timescale not greatly exceeding the age of the Earth. And such a collapse, if it were under way, could be seen as a blueshifting of extragalactic spectra. But in our model the Universe is exploding apart with a violence that gravity has so far been unable to counteract. Out model does not explain the expansion, it merely accepts it as a property conferred on the Universe by its Creator.

The Olbers' paradox too is satisfactorily gainsaid. As we originally conceived the paradox we thought of a Universe filled to plentitude by the radiation of its constituent stars until the temperature everywhere in space approached the several thousand degrees of a typical stellar surface. It is simple to calculate that the time it would take for the observed stars so to light up an initially dark Universe is no less than 10^{24} years, simply because of the large empty volume per luminous star. But the expansion of our model does not allow for the stars to be much older than 10^{10} years. Thus space cannot be at more than a few degrees which we observe. Another cogent objection to their original paradox was not appreciated by the early astronomers. Where was all the glowing energy of their hypothetical sky to come from? Now that we recognize that energy is a conserved quantity that cannot be summoned up from nowhere, we must look for a generating agency. The complete transmutation of all the visible mass in the Universe into light by thermonuclear means falls short of the amount necessary to fulfil the Olbers' paradox a million billion times over.

Actually, the predominant source of radiation in the Universe appears not to come from stars but from the afterglow of the big bang. You will recall that when the Universe began to clear as the free electrons and protons combined to form hydrogen, the material was all at about 3000°. All around us was a hot opaque blanket of gas conforming exactly to the Olbers situation, and we can still see that hot horizon in the very far distance. But if the clearing took place, as it did, ten billion years ago, then the hot horizon as it is presently seen must be situated ten billion light years away. Thanks to the initial expansion, matter at that distance is, and always has been, receding from us almost at the speed of light, and the consequent redshift reduces its apparent temperature by a factor of one thousand to a mere three degrees. That is what Penzias and Wilson detected. So whether radiation comes from stars or is left over from the initial bang, the Olbers paradox is confounded by our model.

That a creature so recently down from the trees should be able to calculate the condition in his Universe back to within a hundredth of a second of its creation must be a source of as much astonishment as pride. Is it not more likely that some or many fundamental aspects have been left to assault our pride in some later and more knowledgeable age? To forestall any complacency we ought to mention one or two sources of unease in the otherwise coherent picture.

To begin with, we know almost nothing of what happened in the first hundredth of a second. Our knowledge of particle physics at such high energies is quite inadequate. We do not even know why the Universe is expanding at all.

Second is the present and ever-preserved disparity between the number of photons and number of baryons in the cosmos. There is but one baryon for every billion or so photons. What a strange and presently inexplicable ratio. There is a remarkable coincidence here too. Had all the photons arisen from the simple transmutation of 100%

initial hydrogen into the presently observed composition of 25% helium by mass and 75% hydrogen, then one billion photons per baryon would indeed have resulted. But in our picture the same number of photons have been present from the very outset, and the disparity in numbers is owed to a more primitive and presently inexplicable cause.

A third great difficulty is the origin of structure in an expanding radiation-filled Universe of the type we envisage. Decades of imaginative theorizing have entirely failed to account for the origin of the galaxies we can see all around us. There is no way we can think of in which they could have evolved from the pristine soup.

Finally there is the greatest puzzle of all, ignoring relatively small-scale features such as galaxies and clusters, the Universe is highly *isotropic*; that is to say it looks remarkably the same in whichever direction we care to look. To a high degree of accuracy there appear to be as many galaxies, as many clusters, as many radio sources, and as many quasars in any one direction as in any other. Even the three degree background which is a delayed picture of the Universe as it was clearing ten billion years ago is extremely isotropic too. This puzzles us because light has a finite speed and nothing, not even information, can move faster than light. That being so, different parts of the Universe that are now simultaneously visible were not causally related for much of the past. How therefore could they have developed in such exactly similar ways? How did one outpost conform in its evolution so precisely to another when both have lain, before this instant, quite outside each other's ken? Something fundamental is surely still missing from our picture. That is not to say it may not be correct in broad outline, but we cannot be certain it is complete.

We can hardly be surprised that some problems remain. The cause for wonder must be that so much is known from observation and explained by current theory. The emptiness and transparency of space have allowed us to survey the contents out to a great distance. Most of the visible mass appears to be locked up in galaxies while most of the radiation is a highly redshifted after-glow from the big bang. Space appears to be very largely empty, for the galaxies are presently about a hundred galactic diameters away from their nearest neighbours, and this intergalactic distance is increasing with the universal expansion. Playing the expansion backwards, we deduce that about 10^{10} years ago the density of matter and radiation must have been very high, a deduction that is confirmed by the isotropic three degree radiation left over, and by the chemical composition of material at the atomic level. The observed graininess of the cosmos is attributed to the universal effect of gravitation. As explained by Einstein, gravitation is a warping of space caused by the presence of local material. But if gravitation has local effects, the mean mass density of the Universe on a much wider scale must also give rise to an overall curvature, a curvature which in turn decides the general geometry and evolution of the whole cosmical structure.

We are at last face-to-face with the central preoccupation of this book. Is the mass density of the Universe sufficient to close up space into a finite volume, or is space infinite? Is space positively or negatively curved? Will the cosmos continue to expand forever or will the gravitational attraction of the nebulae upon each other cause a slowing expansion and eventual recollapse? Do the discernible galaxies contain the majority of mass, or is a great deal more locked up in presently invisible forms? Do we know the masses of even the visible galaxies with any certainty? These are not separate questions, but a single fundamental question posed in a number of alternative ways. In the next chapter we shall find that the geometry of space, the mean mass density of its material contents, and the past and future evolution of the whole metasystem are indissolubly linked via Einstein's field equation. Can we but find a clear-cut answer to any one question, the others will fall at the same time.

Weighing the Universe

WHETHER or not space is curved is, we now recognize, not a question for mathematical debate, but a matter for experimental and observational determination. The minute displacement of starlight as it passes close to the sun indicates that there is indeed a very slight curvature or puckering in the immediate solar neighbourhood. But this observation tells us nothing of the curvature, if any, which exists in the immensities of space between the galaxies. It is that global curvature which determines the grand design of the Universe. One can no more infer the global curvature from solar-system observation than one can deduce the radius of the Earth by triangulation of the front lawn.

In principle, though, a survey of sufficiently distant objects, that is to say, objects whose light-travel time away from us is a significant fraction of the age of the Universe, should lead to a measurement of the underlying geometry. The qualification of 'in principle' needs to be stressed because, with present instruments and techniques, the required measurements are very difficult to make. The difficulties are of two kinds. To begin with, objects at a cosmologically significant distance are mostly so faint that even with the largest telescope it is exceedingly hard to make the interesting measurements with sufficient accuracy. But there is a second and more fundamental difficulty. Nearly all curvature measurements boil down in the end to a comparison of very distant objects like galaxies with similar galaxies nearby, where the assumption is that one is comparing like with like. And such an assumption is generally difficult to justify, for the test-objects, or galaxies, may themselves be evolving as time goes by. If we examine a cosmologically interesting galaxy, i.e. one so distant that the light has taken say twenty per cent of the age of the Universe to reach us, then we are seeing that galaxy as it would have appeared around 2 billion years ago, when it may have been intrinsically brighter or bluer or larger than its counterparts in our vicinity today. So, in comparing far with near we would not necessarily be comparing like with like as we have assumed. For accurate measurements of cosmological curvature we must either understand such evolutionary effects and correct for them or else try to find some class of objects which have not, we believe, been much subject to evolution or change.

Notwithstanding the practical and inherent difficulties, astronomers have devoted considerable effort to the measurement of cosmological curvature. In the summary that follows the reader will encounter a good deal of tangled controversy, uncertainty and downright disagreement. While I beg the reader's patience, I cannot apologize for the rough-hewn vigour of a healthy scientific debate. Although the *product* of science may be an expanding body of knowledge about the Universe, the *practice* of science has to do with the thought-processes and methods of action and argumentation by which such knowledge is won. It is a bright battlefield of clashing egos and ideas which is intensely exciting to its combatants. If the reader gets deafened or confused I have tried to provide a succinct and low-key summary at the very end. But as Schumacher has remarked: 'after all, matters that are beyond doubt are, in a sense dead; they do not constitute a challenge to the living.'

Curvature measurements can be made in three broadly different ways. Firstly, there are direct geometric observations analogous to a surveyor's triangulation of the Earth's surface. Because of the difficulties outlined above, such measurements tend to be inconclusive, though certain constraints can be set. Secondly, we shall see that the past history of the Universe is to some extent tied up with its global geometry. In consequence, there may exist fossil evidence of various kinds bearing on the overall curvature. For instance, the deuterium abundance in our Galaxy may tell of events in past history linked to the finiteness or otherwise of the cosmos, and hence to its curvature. Lastly, we come to observations which bear on the mean mass density of space in the large. According to Einstein, a space totally devoid of matter or energy is necessarily uncurved or flat. Any curvature present in space is a direct consequence, through his field equations, of the material embedded within it. The curvature near the sun can be calculated from the sun's gravitational mass. Likewise, the mean curvature of the Universe should be calculable once the overall density is known. If we can survey a volume of space large enough to be typical of the Universe as a whole and if we can observe and weigh the majority of objects within it, then we can calculate the curvature directly from the inferred density, using Einstein's equations. Our preoccupation in later chapters of the book will be with the mass inventory of the Universe. In this one, we concentrate on curvature measurements made by other means.

Let us begin by asking what is meant by the 'curvature of space.' On the Earthly scale, in which our sense organs have evolved, any curvature there might be is far too small to be of any human consequence. Evolution has not, therefore, equipped us with sense organs capable of registering (or even a brain capable of visualizing) such curvature, and no study of mathematics, however intense, can ever make good a deficiency which handicaps the expert and the layman alike. Even so, the concept of curvature can be rigorously defined and unambiguous methods for measuring it can be agreed upon and put into effect. In this respect, we are all like colour-blind physicists who, while they may be subjectively unable to recognize the difference between azure and emerald, can nevertheless measure and agree upon the different wavelengths of the light waves to which these subjective colours correspond.

If we are incapable of visualizing a three-dimensional curved space, we are thoroughly familiar with the two-dimensional variety, which is simply a curved surface. Since most of the concepts and consequences of curvature carry over naturally from the two- to the three-dimensional variety, we shall follow the usual practice of looking at curvature first in the two-dimensional context, and hope to infer the three-dimensional curvature properties by analogy.

Fig. 5 illustrates five different two-dimensional spaces or surfaces: in (a) a flat plane, in (b) a spherical surface, in (c), in (d) and (e) more irregular surfaces. Consider first the two-dimensional observer O_A in the flat 2-space of (a). If he takes his ruler $O_A P$ and sweeps it completely around, the end P describes the circumference of a circle as shown. Next he can move his ruler and measure the length of that circumference and he will discover the familiar Euclidian result that the circumference has a length $2\pi(= 6.28318\ldots)$ times the radius or length of

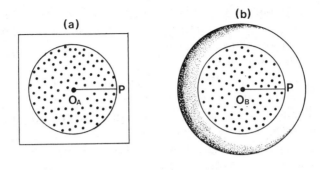

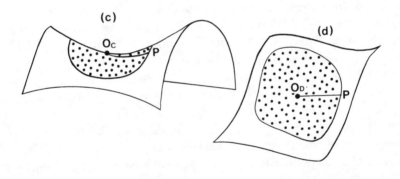

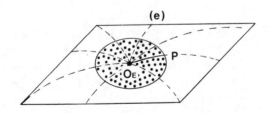

Fig. 5 Different sorts of geometry (see text).

the ruler, which is, say, L. If we pepper his space with equally spaced (on average) dots (the analogy in 2-space of galaxies dotted in 3-space) the number of dots inside the circle will be proportional to the circle's area, which once again we can measure and find to be the familiar πL^2 of Euclidean geometry.

Observer O_B (see (b)) resides on a curved spherical surface. His ruler (which being part of his space is naturally curved with it) likewise describes a circle as shown. This time, however, the circumference will not be found to be 6.28318 times L but rather less in fact, depending on the length of the ruler L in relation to the radius of the sphere. The measured circumferential length of a circle inscribed in a 2-space as compared to its radius can thus be used to infer something about the curvature of the space in which it is inscribed. Moreover, in case (b) there are rather fewer dots within the circle of radius L than there are in (a) for the area of the circle is less than πL^2. Notice also that while the flat space of (a) can be extended indefinitely, in (b) it has a finite extent corresponding to the total surface area of the sphere. For the observer (an ant, say, on the skin of an orange) confined to the 2-space (b) the space has no boundary, and yet as judged by the total number of dots within it, the observer will find it finite in extent. This raises the possibility of a curved 3-space which is finite in *volume* but which likewise has no boundaries or indeed any preferred 'centre.'

But not all curved 2-spaces are necessarily finite, as can be seen in the saddle-case (c). Here the circumference of a circle is actually greater then 6.28318 times L while the area of the circle, and hence the number of dots within it, is greater than for a circle of equivalent radius drawn in the flat-space (a). So, although the space can be inferred to have curvature, in the sense that it is non-Euclidean, the space is also non-finite because the saddle-surface does not close upon itself but can be continued indefinitely.

The 2-spaces (a), (b) and (c) are all highly regular and symmetric. For instance, the view as seen by the observer in each is the same in whatever direction he cares to look. Such special spaces, in which there is no preferred directions, are called *isotropic*. Space (d), however, with all its lumps and bumps and wrinkles, is definitely *not* isotropic; even if the number of dots per unit area in this space is everywhere the same, the observer will see more of them crowded together in some directions that in others.

The spaces (a), (b) and (c) also enjoy a second and rather special symmetry. They are *homogeneous*, that is to say, wherever the observer is situated within them, the geometrical properties will appear to be the same; no one part of them is geometrically distinct from any other, so that the properties of each whole space can be inferred unambiguously from the local properties of any one part of it. That homogeneity and isotropy are distinctly different symmetries can be seen by examining space (e). To an observer O_E situated in the centre of the dimple, space in all directions appears to be equivalent, so it is isotropic. But the space is not homogeneous, because the geometry of it will appear different to observers situated at different sites, e.g. inside or outside the dimple.

Physical space is of course three-dimensional but most of the concepts and arguments outlined above for 2-spaces carry through quite naturally. For instance, an isotropic 3-space is one which appears to be geometrically identical in all directions, while a homogeneous 3-space is one whose geometrical properties appear the same to observers situated anywhere within it. But now the dots are replaced by galaxies, the circles by spherical surfaces equidistant in all directions, and the areas of circles by the volumes of those spheres. Just as in a closed 2-space, the area of a circle (and hence the number of dots within it) is less than the area of a circle of equivalent radius drawn in a flat 2-space, so in a closed 3-space the volume of a sphere (and hence the number of galaxies within it) is less than the volume of a sphere of equivalent radius drawn in flat or Euclidean 3-space. Likewise, there are curved but open 3-spaces in which the volume (or galaxy population) increases more rapidly with distance than in a flat-space. Counting galaxies as a function of their distance away from us might, in principle at least, indicate whether the actual Universe of Space in which we live is open or closed.

But we are jumping the gun. What if the actual Universe is lumpy and bumpy with a curvature that varies from place to place so that nothing can be inferred of its global properties from the geometry nearby? In principle, the geometry could be highly complex with all manner of bizarre and wondrous properties.

Fortunately, all the observations concur in suggesting that on the large scale the Universe is highly isotropic. There appear to be as many faint galaxies and radio sources in one direction as there are in any other. Even more persuasively, the cosmic big-bang radiation, which comes from an enormous distance away (where the redshift is no less than a thousand), is observed to be isotropic to around one part in a thousand, and this effectively rules out a bizarrely curved Universe. There may be tiny wrinkles here and there in the vicinity of concentration of matter, for instance close to the sun, but on the global scale these wrinkles appear to be insignificant.

But if the Universe is isotropic, is it homogeneous as well? Here we are not on such observationally firm ground: for instance, we could be situated at the centre of a local dimple as is the observer in Fig. 5(e) above. But we usually ignore this possibility on probabilistic grounds. We can already photograph a billion or more isotropically distributed galaxies not so very different from our own. Is it likely that out of all these millions ours just happens to be situated in the very centre of a special dimple? If it were not so centred the Universe would appear non-isotropic, which we know is not the case.

For the above observational reasons, cosmologists start rather confidently from the assumption, grandly entitled 'the cosmological principle,' that the Universe is everywhere homogeneous and isotropic. And this principle leads to enormous simplifications. To begin with, it implies that the global geometry can be inferred from the geometry within an observable range. This brings in cosmology from the wilds of theology to the fold of observational science. Secondly, the cosmological principle drastically restricts the class of geometries which can obtain in the real world. In fact, there are only three geometries in 3-space which are both isotropic and homogeneous. One is the curved but closed 3-dimensional analogue of the 2-space geometry on the surface of the sphere (see Fig. 5(a) above), the other is the curved but open analogue of the saddle-geometry in 2-space (Fig. 5(c)). The third space is ordinary Euclidean. A closed geometry calls for a finite Universe, while the other two are infinite. They differ in particular in the amounts of volume they contain within a fixed range of space. Therefore, it should be possible, in principle at least, to ascertain the geometry of the actual world by counting galaxies as a function of distance, or by some analogous technique.

One might choose to regard the curvature of space as God-given, and therefore not in need of any more primitive cause. However, on grounds of economy-of-hypothesis, cosmologists usually allow God no arbitrary role in this. They note, following Einstein, that the best theory of gravitation we have appears to be the general theory of relativity. In this they assume gravitation is nothing other than spatial curvature produced by local mass. And since the Universe is filled with material galaxies whose gravitation also will produce a component, it is most economical to assume that the entire curvature of the Universe is a consequence of its material contents.

This further assumption, which goes beyond the cosmological principal, leads via some not very difficult sums to a momentous consequence. There turns out to be a unique and simple mathematical relation between the curvature of space, the mass density ρ of the material it contains, and its expansion rate. Once you have any two of the three quantities, curvature, expansion rate and mass density, the third can be calculated at once from the Einstein equations.

We do not need to be mathematicians to appreciate this simple Einstein equation. Einstein says there is a critical density ρ_c calculable from the expansion age T_0 of the Universe by:

$$\rho_c = \frac{3}{8\pi G T_0{}^2} = 1 \text{ atom per cubic metre}$$

and if the actual mass density of space is greater than the critical value then the Universe is closed and finite (G is simply a constant denoting the strength of gravitation). If it is less than the critical value the Universe is curved but open and infinite.

So, what is the answer? Hopefully the observed density will turn out to be so much larger or smaller than the critical value that no doubt will remain as to the conclusion.

Observations of the distances and redshifts of galaxies lead to an expansion age T_0 within a factor 2 either way of 10^{10} years. The critical density then works out to be:

$$\rho_c \sim 2 \times 10^{-29} \text{ gms/cc or about 1 atom per cubic metre}$$

A very rough idea of the minimum *actual* density can be gained by assuming that most matter is collected into visible galaxies, then counting the number and estimating the masses of all the galaxies in a representative volume of space. The best estimates then yield:

$$\text{Density observed} \sim 2 \times 10^{-31} \text{ gms/cc}$$

which is only about one per cent of the critical density needed for closure of the Universe.

Naively we might suppose that this simple calculation has settled the question once and for all and that the Universe is open and infinite. Experience, however, shows that naivety seldom pays in astronomy; we should look at some of the caveats. To begin with, our so-called 'observed' density is very much a lower limit to the true value. For instance, it may not be true, as we have assumed, that most material is gathered into visible galaxies, or indeed visible anythings. There may be invisible galaxies or other denizens of space presently unknown, in which the remaining matter is locked. And then again, as we shall see later, the masses of galaxies are highly uncertain with clues which point to masses ten times higher than we have here assumed. Moreover, there are entities such as neutrinos, radiation, antimatter, intergalactic gas and gravitational waves which can all contribute to an increased actual density.

In fact, the miracle must be not that the calculated density is so much less than the critical value but that it lies so close. After all, astronomers are used to huge numbers, and it would not have surprised them to find an observed density either millions of times larger or smaller than the critical value. On the contrary, they are intrigued by how close the two densities, one calculated, the other inferred from observation, appear to be.

Is it merely a coincidence or does it suggest that space is indeed closed but our primitive observations have *so far* located only a per cent of the total contents? We shall return to such questions later in much more detail. For the moment, we merely emphasize the close connection between the curvature and mass density of space. If we can measure the density we can infer the curvature or vice versa. The crucial quantity is the true density divided by the critical value. This quantity will occur so often that to save space we shall simply follow normal practice and call it Ω (omega). If Ω is more than 1 the density is finite and closed. However, if Ω is less than 1, the density is greater than the critical value and the Universe is below its critical value implying an open, infinite Universe. The value of Ω, crudely and naively inferred from the visible galaxies alone, which is clearly a lower limit, is about 0.01.

Although in Einstein's theory curvature and gravitation are synonymous, on the mechanical level gravitation manifests itself as a force of attraction between any piece of mass and its neighbours. In our expanding Universe, where the visible matter is largely gathered into galaxies at enormous distances from one another, galaxies which are in any case flying apart at speeds comparable to light, one might suspect that the weak force of gravity would scarcely slow the expansion rate. To underline this suspicion, note that the deceleration of a receding galaxy due to the gravitational attraction of its nearest neighbour is one thousand billion times less than the familiar gravitational acceleration at the Earth's surface. Gravity would seem as impotent to halt the expansion as an elastic band to slow down a supertanker at sea.

But this is not necessarily so. To begin with, gravity appears to be an elastic band that may weaken but does not snap, no matter how far apart the galaxies between which it is stretched. And although the resulting decelerations are very small indeed, they have remorseless aeons of time over which to act. A rather simple calculation, based upon Einstein's theory of gravity, shows that gravity can indeed halt the expansion, and turn it into a collapse, provided only that the average density of matter in space today exceeds a certain critical value, which depends upon the expansion rate. Obviously the faster the expansion, the greater the density and hence the gravitational attraction, required to reverse it.

It turns out that the critical density is the very same density which appeared earlier (ρ_c) as that needed to close the Universe. Thus the dynamical fate and the geometrical character of the Universe are indissolubly linked, with both controlled by the average density of matter. Both the spatial and temporal extent of the Universe are controlled by a single number Ω (omega) which is the ratio of the actual density of space to a critical value calculable from the expansion rate. If Ω is more than one we live in a spatially finite Universe where expansion will one day cease and turn into an irreversible collapse with catastrophic consequences. On the other hand, if Ω is less than one, the expansion will continue indefinitely and we live in a Universe without end. No wonder astronomers are anxious to determine Ω and to settle these momentous issues once and for all.

We might pause for a moment to look at the two alternative fates reserved for us, depending on the value of Ω. If Ω is 2, say, implying a closed Universe, nothing very dramatic will happen for a few billion years. The Universe will continue to expand and the big bang radiation will continue to cool below its present 3^8, though astronomers will note that the rate of recession of galaxies is gradually slowing down. In ten billion years, the recession will halt when the radiation temperature is about $1.5°$ and thereafter galaxies will increasingly exhibit blueshifts as they begin to fall back in towards us. A few billion years after that, neighbouring galaxies will begin to collide with our own, showering the sky with red giant stars from distant parts. The big-bang radiation temperature will now be rising rapidly as the radiation is compressed, until the sky temperature reached $300°$ or more, even at night, and life will become insufferable. As galaxies crowd even closer together and commingle violently the gravitational forces increase, rapidly accelerating the whole collapse process. Soon, the radiation of space will become hotter than the stellar surfaces temperature and stars will begin to sweat and burst. At even higher temperatures radiation will knock all electrons from atoms, then all protons out of nuclei until all the present structure of the Universe is redissolved into the fiercely hot soup from which it originally emerged. It is as if a movie film has been taken of an expanding Universe once then run backwards. No living thing could possibly survive the final dive into the furnace; even familiar structures right down to the atomic level will be obliterated and no record will remain to tell that humans ever existed, or what they were here for.

It is a pretty grim scenario all round, though some unanswered questions remain. For instance, can such a Universe burst forth, and go through its whole cycle again and again; and will each cycle repeat precisely the course of its predecessors? For the moment at least physics is powerless to tell us. Some astronomers have even doubted that a Universe can apparently collapse at all; they have argued that the arrow or direction of time is determined by the expansion and that if a collapse set in the arrow would reverse, convincing the inhabitants that it continued to expand. But these speculations are not generally considered respectable.

An open Universe with Ω less than one will suffer no such dramatic fate. The expansion will continue indefinitely so that even our neighbour galaxies will become very distant and faint. But surprisingly the sky will not empty altogether, for the total number of galaxies we can see will actually *increase*. This comes about because sufficient time will have elapsed for the light from galaxies beyond our present horizon to reach us for the first time.

The ever-expanding Universe will, however, cool with the big bang temperature falling towards zero. Stars will all exhaust their fuel and collapse in ten billion years to become non-luminous black dwarfs, neutron stars or black holes. Despite their lack of radiation these stellar remnants will interact with one another gravitationally by random encounters. Most will be hurled into the ever greater darkness beyond the galaxy while the remainder, in the course of a billion, billion, billion years, will gather into a single massive black hole at the centre. Darkness and cold unimaginable will settle upon the infinite voids of space, but physical processes will not cease altogether. Every eternity or so, bizarre quantum processes will stir the embers of white dwarfs into sparks which feebly and faintly illumine the void.

As all the heat and energy sources give out, life will become difficult for even the most competent of species. Surprisingly, however, there would appear to be strategies for indefinite survival. With the falling temperature, life itself must cool and slow down so that metabolic rates fall to a billion billionth of their current values. Organisms must become so sluggish that even the blinking of an eyelid will occupy an aeon of time. Not much fun, you might think, but will it matter? If thought processes are slowed, as they will be, to a correspondingly sluggish rate, life will seem just as exciting to its icy inhabitants, for their Universe will provide an inexhaustible benefit of the one thing they will need, namely Time.

So Ω controls all: the finiteness of space, the dynamics of matter and the destiny of sentient life. But although it determines the future, apparently it has not left its mark so dramatically upon the past. If only it had, we might expect to find clear fossil evidence testifying to its value. It appears, however, that the earlier stages of a closed and open Universe do not differ so very much from one another. Nevertheless, there is hope that some fossil clue remains to be unearthed. For instance, if Ω is more than one, then expansion is decelerating significantly, and the expansion rate, as judged from the spectra of very distant, and therefore ancient, galaxies, should appear distinguishably larger than the present rate nearby.

Before we examine the observational evidence, let us review the theoretical situation as it stands now. To begin with, we have recognized that on the large scale, space may be curved, and whether or not it is, and in what way, must be determined by observations. Since the Universe would appear to be both isotropic and homogeneous, the permissible geometries or curvature it can have are essentially limited to two: open and infinite or closed and finite. Since curvature and gravitation are synonymous, we believe the curvature is caused by the gravitational effect of the enclosed matter. According to Einstein's theory, the Universe will be closed if its density at present exceeds the critical value of about 1 atom per cubic metre. The density also controls the dynamical fate of the Universe; it will recollapse if the same critical density of 1 hydrogen atom per cubic metre today is exceeded. Thus Ω, the actual density divided by the critical value, determines both the finiteness and fate of the cosmos. If Ω exceeds 1 the Universe is finite and doomed to recollapse into the fireball; if Ω is less than 1 space is unlimited and time will be without end. An estimate for Ω which assumes that the whole density is comprised in the visible galaxies yields $\Omega = 0.01$. This is certainly less than 1, but not by so much that the presently invisible material, of which there certainly must be some, could not account for the remaining 0.99. This remaining material, which may or may not exist, is usually referred to as 'the missing mass.'

Before we go in direct search of the 'missing mass' we shall, in the remainder of this chapter, review the global evidence on Ω. Do observations of the global geometry, the global dynamics or the global history set useful constraints on Ω, and hence upon the total proportion of the hidden mass? Such observations cannot of course be of space itself, but only of its visible inhabitants the galaxies. We need to look at galaxies whose light-travel-times away from us are a significant fraction of the age of the Universe, that is to say, very distant galaxies at high redshift. Distances so large, expressed in kilometres, are virtually meaningless; that is why we speak either in terms of redshift or of light-travel-time expressed as a fraction of the age of the Universe. The relation between these two interchangeable measures of cosmic distance is shown in Fig. 6 for reference for different models of the Universe.

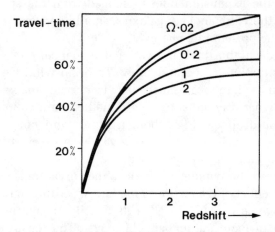

Fig. 6 Light travel-time from highly redshifted objects for four different models of the Universe. Travel-time is given as a percentage of the age of the Universe and shows how long light takes to reach us from an object with a given redshift. (Adapted from P. E. Peebles, *Physical Cosmology*, Princeton University Press, 1972, p. 187.)

It turns out that for cosmologically significant observations, we generally need to look backwards 30% of the age of the Universe, which corresponds (see Fig. 6) to redshifts of more than 0.5. Even the brightest galaxies at such a range appear as no more than fuzzy dots hardly distinguishable from faint foreground stars even on the deepest photographic plates. With objects so faint it is such a struggle, even with the largest telescopes, to measure their brightnesses, their colours, their diameters, and in particular their redshifts, that for only a handful is the necessary information complete.

We begin with the famous 'standard candle' test which is predicted on the assumption that a class of galaxies can be discovered all wih the same *intrinsic* luminosity. If they can be found, a more distant (i.e. higher redshift) member of the class should *appear* fainter than nearby (low redshift) classmembers by an amount which depends not only on the redshift-difference between them but also upon the global curvature. To see why curvature enters, we return as ususal to our curved 2-space analogue.

In Fig. 7 a 2-dimensional galaxy G emits 2-dimensional light waves, signified by the little arrows in all directions on the 2-dimensional space or surface. The apparent brightness of G, as judged by the two-dimensional observer O, at a distance R from G, depends on the number of arrows per sec hitting the fixed diameter D of his eye. The number N so hitting is the fraction D/C(R) of all those emitted, where C(R) is the length of the total circumference at R. But, as we argued earlier, the dependence of the circumference C(R) upon the distance R is a function of the curvature of the surface upon which the circle is inscribed. On a flat surface $C(R) = 2\pi R$ and so the apparent brightness $N = D/2\pi R$. But on a closed (spherical) surface, we argued that C(R) is less than $2\pi R$ so the apparent luminosity $N = D/C(R)$ will be *greater* than in the flat space. Conversely, on the saddle-surface where C(R) is greater than $2\pi R$ then the apparent brightness $N = D/C(R)$ will be *less* than in the flat case. Shifting upwards to 3 dimensions the same argument as always carries through. We conclude that in a closed Universe galaxies at a fixed distance (redshift) will appear brighter than they would in Euclidean space, and in an open Universe they will appear fainter. Observations of the apparent luminosities of galaxies as a func-

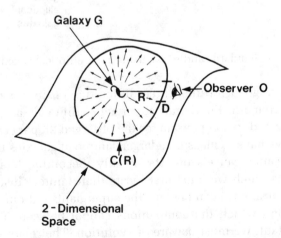

Fig. 7 Brightness in a curved space.

tion of their redshifts can therefore disclose the global geometry. All this supposes, of course, that the galaxies one is comparing both far and near, have the same intrinsic luminosities; that they are, so to speak, standard candles.

Can improbably standard candle galaxies be found? If we examine *nearby* clusters of galaxies, we often find, close to the centre, a supergiant elliptical very much brighter than all its companions (see Fig. 8). And these supergiants, for completely mysterious reasons, all seem to have the same luminosity within twenty-five per cent or so. This is a double piece of unexpected good fortune because such galaxies are (a) easy to identify as such because of the large clusters around them and (b) the most luminous galaxies known and therefore the easiest to observe at the great distances at which we have to work.

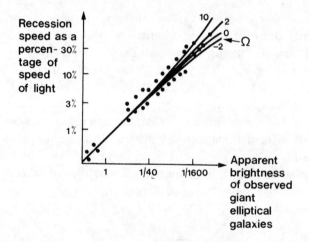

Fig. 8 Standard candle test, using giant elliptical galaxies.

Despite this good fortune, three difficulties, in increasing order of importance, remain. Firstly, all supergiant ellipticals are not *exactly* co-luminous and to compensate for the observed 25 per cent scatter in brightness we have to measure a larger sample of galaxies in which we hope this scatter cancels out statistically. Secondly, to identify the supergiants as such we must first locate their fainter cluster companions which means we can observe the supergiants out only to redshift of 0.5, beyond which the companions cannot be seen. Thirdly, and most important, we must beware of *evolution*. The relative constancy of our candle galaxies has been arrived at by comparing one supergiant with another in relatively *nearby* clusters where we can neglect the curvature effects as insignificant. But all we can strictly infer from this is that supergiant ellipticals of roughly the *same age* have the same luminosity. We cannot go a stage further and prove, as we really need to do for the test, that supergiant ellipticals of *all* ages have the same luminosity. Because of light-travel-time the more distant galaxies in our sample will necessarily be seen when younger than those nearby, and if the intrinsic luminosity of galaxies evolves with time, as seems quite likely, this vitiates our test. For instance, if galaxies were brighter in the past, then the more distant galaxies in our sample will appear brighter than they should be, and this will mimic precisely the appearance of a closed Universe with no evolution. So, unfortunately, the effects of curvature and evolution may be intertwined in the observations.

Neglecting evolution for now, Fig. 8 shows the latest results of the standard candle test, results obtained largely by Allan Sandage working with the Hale 200-inch telescope over the past thirty years. Each of the dots denotes a separate program galaxy, while the curves show what is to be expected if the Universe is closed, flat, open or in a steady state (see later). On the face of it, the observations marginally support a closed Universe, but the redshift range over which measurements have been obtained is too small to be conclusive. From these measurements, and ignoring evolutionary effects, Sandage could only say that to within 50 per cent probability Ω lies in the range 0 to 2.

But what of evolution? How could it and how does it affect the outcome of the standard candle test? If galaxies were more luminous in the past, the more distant ones will appear brighter than they would otherwise be, mimicking the effect of a closed Universe and so falsely increasing the measured value of Ω. Conversely, if galaxies brighten with time, Sandage's measure is too low and must be increased. The question is, do galaxies brighten or dim with age, and by how much?

The luminosity of a galaxy is the sum total of light from its myriad diverse stars. To compute evolution, one needs knowledge of the detailed stellar population and stellar age-structure of the galaxy, together with a fair idea of what happens to the individual stars as a function of their ages, masses and chemical compositions. So it is complicated, and doubly so because there is no really nearby giant elliptical to give an insight on the likely stellar population. Opinions on evolution have see-sawed enough over the past decade to warn us against over-dogmatism.

The present COWDUNG ('Conventional Wisdom of the Dominant

Group') has it that elliptical galaxies are fading with time, so that nearby specimens are about 20% dimmer than the most distant objects in Sandage's sample. Sandage's $\Omega = 1 \pm 1$ is thus estimated to be too high by as much as one. The argument goes like this. Elliptical galaxies are all very red and therefore very old (the young blue stars having become extinct). Some of their red light must be coming from billions of faint red dwarfs and some from a much smaller number of super-luminous red giant stars. Modern infrared measurements suggest that the giant contribution completely overwhelms all the dwarfs and that therefore the luminosity evolution of the galaxy as a whole will be controlled only by the giants. Fortunately, all red giants, irrespective of age, are thought to evolve in the same way. Galactic evolution depends, therefore, only on the changing number of giants in a galaxy as it ages.

We recall from Chapter 3 that red giants are merely dying dwarfs enjoying a last spectacular fling. Thus, the luminous output of an elliptical is thought to be a reflection of the death-rate among its dwarfs which, so it turns out, can probably be estimated within reasonable limits. The upshot of all this is that ellipticals are probably fading with time. Our standard candles are, therefore, not as standard as we fondly assumed and to compensate for the suspected evolution Sandage's figure for Ω must be revised downward to around 0 ± 1 which is a result so equivocal and so uncertain as to be uninteresting. All it does with any certainty is rule out a very dense Universe containing considerably more mass than the minimum required to close it. In view of the heroic amounts of work and of valuable telescope time absorbed in reaching it, this is a disappointingly limp conclusion.

If we could be certain that ellipticals were dimming with age, we could at least assume that Ω, as measured by the standard candle test, is an upper limit to the true value; i.e. we could set an upper limit to the universal density. Recently, however, theoretical astronomers, with the aid of computers, have been reappraising the effects of galaxy-galaxy collisions. It now seems likely that giant galaxies in the centre of large clusters, i.e. the very beasts we have chosen as our standards, may be able to capture and engorge their lesser companions during close encounters. This dynamical effect could actually increase their luminosities with age, so the whole business of evolution is back in the melting pot once again.

A second important method of search for curvature is the standard ruler test. As a stick of fixed physical length is moved further away from us its apparent size, that is to say the angle it subtends at the eye, will naturally fall. If, in a Euclidean Universe, you double the distance then you halve the angle. In a curved Universe, things are not so simple. To see why, return as usual to our two-dimensional curved space i.e. to the geometry of a two-dimensional analogue, (i.e. surface), inscribed on a sphere. Once again, the apparent angular size of a rigid stick will be the angle between the light rays arriving at the eye from opposite ends of the stick. In Euclidean space, the rays travel in straight lines, where a straight line is the shortest path between two points. In curved space, the rays will still travel by the shortest conceivable paths, but these paths will no longer be straight. For instance (see Fig. 9) in the two-space inscribed on a sphere the light will travel along the great circles followed by pieces of string stretched taut between the ends of the stick and the eye. A little thought, or better still some fiddling with a match-stick, pins and cotton on the surface of a grapefruit, will soon convince you that as the stick recedes, the apparent angular size falls until it reaches a minimum and therefore it will *increase*. And this apparently paradoxical result can carry through into curved three-space where distant galaxies might actually appear larger than identical objects nearer by. Fig. 10 illustrates the physical reason for this. The gravitational attraction of intervening material enclosed within the light beam curves the rays OG_1 and OG_2 inward so that they appear to come not from the actual galaxy G_1G_2 but from the larger

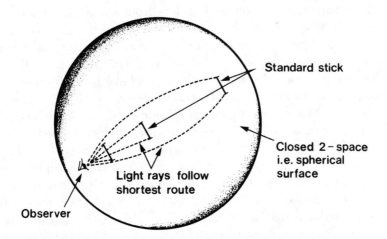

Fig. 9 The standard stick test in a closed or finite Universe. The light rays follow the shortest paths, which are now curved. The apparent angular size of the stick first decreases and then increases as the stick is removed to greater and greater distances.

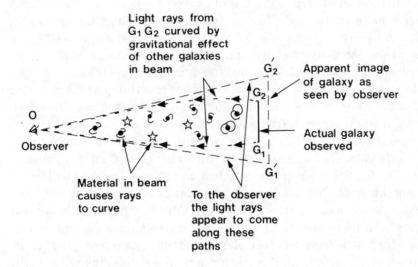

Fig. 10 The magnification of distant images due to gravitational curvature of material inside the light beams.

image $G'_1G'_2$. This magnification of distant images will occur in any Universe with gravitational curvature, but it will be obviously more pronounced in a denser (i.e. closed) model.

Unfortunately, this method suffers from some of the same weaknesses inherent in the standard candle test. Firstly, standard sticks have to be found; secondly, the angles involved may be very small; thirdly, we must beware of evolution. Nevertheless, several astronomers have proposed standard sticks and have tried to measure Ω by plotting the angular sizes of various classes of objects against their redshifts. Once again, supergiant elliptical galaxies have been chosen as likely standards and Fig. 11 shows the results obtained by William Baum at Flagstaff, Arizona, where he has measured their apparent diameters as a function of redshift and found Ω to be in the probable range 0.6 ± 0.4, a result which, he argues, is independent of evolution and is not consistent with a Euclidean Universe. But the measurements are so delicate that we can argue, on the contrary, that they must depend on the history of the objects whose luminosity is probably changing and which, in any case, may be ingesting their nearby companions and swelling as a result. The much larger diameters of double radio-sources have likewise been used by George Miley in Leiden, but the scatter in the diameters is so large that any value of Ω between $+1$ and -2 can be reconciled with them.

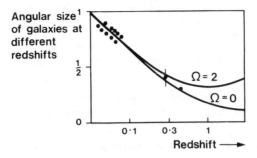

Fig. 11 Angular size of giant elliptical galaxies against redshift (distance) for different cosmological models.

The unfortunate interrelation between geometry and evolution in the two curvature tests, together with our uncertainties about the amount of galaxy evolution, means that, as they stand, the two direct curvature tests above tell us very little definite about Ω. This raises the alternative question: can tests be devised to measure evolution independently of curvature? For if they can, then we might hope to use the solution so found to correct the geometrical tests above, and hence arrive at a more accurate value for Ω. Somewhat surprisingly the answer turns out to be 'yes,' for the mathematics of curved space shows that evolution can be disentangled from curvature more easily than vice versa. Let us therefore digress for a moment to examine what is known of evolution.

Two separate questions have been asked about cosmological evolution. Firstly, is there any evidence whatever that the Universe is evolving? Secondly, in particular what can we observe of evolution amongst the giant elliptical galaxies which play such a crucial role in the curvature determinations?

To begin with the first question, one might suppose that there can be doubt as to the answer. By its very nature an expanding Universe is evolving from a dense to a less dense state. But this is not necessarily so, for if we were prepared to admit that new galaxies could somehow form out of the vacuum to replace those carried off by expansion, it is possible to imagine a Universe which, in its gross properties like density, is unchanging with time. If the notion of conjuring atoms out of the void is hard to take, remember that the alternative of forming them all in a single flash at the big bang is equally implausible. Indeed, the big bang theory is felt by some to be scientifically unpleasing precisely because it sweeps so much under the carpet. Difficulties like isotropy, the proton-photon ratio, the rate of expansion and galaxy formation are all referred backward to the unobservable and therefore scientifically unreachable epoch of the initial bang, and some astronomers believe this smacks of theology.

To meet this objection, three young physicists up at Cambridge in the 1940s proposed an attractive model called the Steady State Theory. Hermann Bondi, Tommy Gold and Fred Hoyle pointed out that in one sense the cosmological principle was incomplete. While observers dotted throughout *space* are presumed to be equivalent, in that they all are supposed to get the same large-scale view of space, the same symmetry does not extend to the time dimension. Observers born earlier on see a hotter, denser cosmos about them than their descendants because the whole Universe sprang into existence a mere ten billion years ago. To overcome these inhomogeneities and naiveties the three young Turks proposed 'The Perfect Cosmological Principle' which asserts that the Universe will appear broadly identical from all positions of space *and at all epochs of time*. Theirs is a Universe radically different from Einstein's; without a beginning, without evolution and without an end. It is infinite and although it expands, new matter is spontaneously and continuously created to replace that which is expanding over the horizon.

Not only is the steady state theory mathematically pleasing, but it is a strong theory which makes clear and firm predictions that can be subject to observational test. For instance, there can be no evolution *by definition*. The density is and will always remain equal to the critical value 10^{-29} gms/cc. The mathematics is necessarily different from the Einstein matter-conserving mathematics and Ω becomes a purely geometrical factor unrelated to the density. For the curvature tests outlined above, the theory predicts a value for Ω of -2, which is why we draw in $\Omega = -2$ lines in versions of the graphs. Since galaxies are continually in creation, according to the theory they can exhibit all ages from the infinitely old to the very young.

The simplest observational test for evolution is called a 'number-count' which works like this. We take a deep photographic plate of a region of sky and identify all the galaxies on it. There will be a few bright nearby galaxies, many more dimmer ones seen in the distance, with the largest number of extremely faint objects at the limit of detectability. If we count and compare the number of galaxies in each range of apparent brightness we are, in effect, sampling the population density of galaxies as it was at various epochs in the past and comparing it with the present density. Geometrical effects depending on Ω cancel out to first order so the test is insensitive to Ω, but sensitive both to expansion and evolution. Expansion will decrease the comparative numbers of fainter objects by redshifting and hence diluting their signals, whilst evolution, if it works in the sense that galaxies were, as we have argued, intrinsically brighter in the past, will act oppositely to increase the comparative numbers at the faint end.

The number count is an ancient test first devised by Herschel in the eighteenth century and used by him to argue that the galaxy is an island of stars of finite extent. It is *relatively* simple to carry out, for it requires no time-consuming redshift observations and it can be used on stars, on galaxies, on radio, X-ray or infrared sources or indeed on any homogenous class of objects distributed throughout space.

We are specially interested in giant elliptical galaxies, and modern photographic emulsions enable a large telescope at a dark site to pick up such galaxies out to a redshift of about 1 when we expect evolution to become appreciable. Several groups have recently reported number counts from such plates and these marginally suggest some evolution, but the evidence is not persuasive and certainly not accurate enough to correct the curvature and so reach an unambiguous value for Ω. In fact, the results are too vague even to rule out the steady state theory.

The radio number-count, however, tells a dramatically different story. A radio astronomer can search the sky and, just like his optical counterparts, produce the comparative number-count for sources of different strengths. Some powerful radio galaxies and radio quasars, which dominate the surveys, may emit as much radio as light energy and so the much bigger radio telescope can pick them up at redshifts more than the value of 1 to which the optical instrument is presently confined. Thus a radio number-count should contain some cosmologically significant information on radio-source evolution.

Radio source-counts have been in progress for twenty years or more and it is now generally agreed that they reveal clear signs of dramatic evolution among radio sources. This result has, however, not been reached without a great deal of uncertainty and passionate controversy along the route. The earliest surveys pointing towards evolution were subsequently shown to be in error because inadequate care had been taken to account for 'confusion,' that is to say for the presence of more than one source at a time in the telescope beam. Different observatories were reporting different results and even today surveys carried out at different radio wavelengths are not in exact agreement. Longer wavelength surveys generally reveal a much more dramatic evolution than short wavelength observations, while the number of bright sources, against which the faint ones must be compared, is so small that there is some room for statistical uncertainty.

Even so, all but the most unrepentant steady staters, like Sir Fred Hoyle, now concede that the radio-counts bear witness to a surprisingly dramatic evolution among radio sources. In fact, a galaxy today is only one thousandth as likely to be a strong long-wavelength radio source as its counterpart at a redshift of 1.

Quasars, because they have very high redshifts in the range 0.1 to 3.5, should be ideal for cosmological studies. Unfortunately, they have too large a scatter in their intrinsic luminosities and sizes to be useful in the standard curvature tests; even worse, individual quasars can vary in luminosity by factors of a hundred on a timescale of weeks. Even so, number counts of quasars can be measured and these agree with the radio counts in implying strong evolution. For instance, if we count 30 quasars above one brightness level of x units on a certain plate then if we recount down to a level of half x we expect to find 80 quasars in an unevolving Universe, whereas we actually find about 150. Quasars were significantly more common in the past than they are today and this is borne out by their redshift distribution, which suggests that the quasar phenomenon, whatever it is, was far more prevalent at redshifts of 2 than it was either before or afterwards.

Such observations are very hard to reconcile with the steady state theory, which also has great difficulty in accommodating the observed radiation background discovered first by Penzias and Wilson. Since a steady state Universe can never have undergone a superdense phase, we cannot appeal to a big bang to supply the 3-degree background. Instead, Hoyle has proposed that the background is emitted by needle-like dust-grains relatively nearby in intergalactic space. But if that is so, the observed isotropy is difficult to understand because the nearby galaxies are certainly not isotropically distributed, being gathered into irregular clumps and clusters, and if that is so, why should the dust behave differently?

For the above reasons, the steady state theory has, in the minds of most astronomers, to be discarded in favour of the more conventional big bang models. There appears to be strong evolution among quasars and radio galaxies; indeed the evolution is so dramatic that it is difficult to explain. But for giant ellipticals evolution is very indeterminate so that we still do not know what corrections to apply to the standard stick and standard candle curvature tests, and as a result they still do not yield any firm results for Ω. We must therefore look for evidence elsewhere.

The most accurate quantity we can measure for a galaxy is its line of sight velocity. Optical telescopes can measure galactic velocities spectrographically out to a redshift of about 0.5 (recession velocity of order of 100,000 km/sec) with an uncertainty of 100 km/sec which corresponds to an error of only about a tenth of a per cent. Radio telescopes can measure velocities with an uncertainty as low as 10 km/sec, but only out to redshifts around 10,000 km/sec. Such a precision contrasts with the thirty or forty per cent uncertainties that persist in the measurements of galactic luminosities and diameters, and suggests we look for purely dynamical tests of Ω based largely on these more accurate velocity measurements. There is much active work in this field at present and we shall now review the results.

The velocity of a galaxy can be thought of as the sum of two components. Firstly, there is the recession component which depends on a galaxy's distance away from us and the rate of expansion of the whole Universe. As we have seen, such recession speeds can be very large and we expect all galaxies located in the same region of space to be receding at the same speed away from us. But galaxies are not rigidly locked to the expanding space in which they find themselves, for observation shows that galaxies have individual or random velocities of their own amounting to a few hundred km/sec; velocities which are large enough to be measured but too small to disturb grossly the picture of a smooth expanding flow.

From where do these random velocities come? One can think of two possible origins. They might be primeval, that is to say, remnants of a once turbulent flow in the gas from which they formed. But this is most unlikely for the following reasons (see Fig. 12). If one among a group of galaxies has a high primeval velocity with respect to its companions, then it will set off in the direction of another group which, because of the general expansion, is receding away from it. When, at a later time, it reaches that second group, its velocity *relative to that second group* will be lower by the relative recession speed of the two groups. So, as time goes by, galaxies gradually settle among companions with a congenial velocity and any early distribution of random relative velocities gradually dies away. From such a simple argument one can show that a huge random velocity of 100,000 km/sec at an epoch of redshift 1000 will have fallen by a thousand times to a mere random velocity of 100 km/sec today. To explain as primeval the presently observed random velocities of perhaps two or three hundred kilometres per second one would need to presuppose primeval velocities close to the speed of

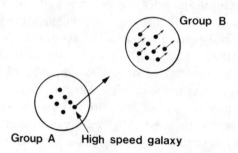

Group B

Group A   High speed galaxy

Fig. 12 Galaxy speeds in an expanding Universe. If a galaxy has a high speed relative to the companions in its group (A) it will leave the group and travel towards another group (B). Because of the expansion of the Universe, however, B will be receding from A. Thus when the galaxy arrives at B it will have a lower velocity relative to the B group members than to its original companions in A. It will carry on until it settles permanently within a group with congenial velocities. In this way the random velocities of galaxies relative to their companions decay with time.

light, velocities which could not have been maintained in a pre-galactic gas.

Therefore, one resorts to a second and more reasonable explanation. Galaxies, we know, are not smoothly distributed in space, but are clustered and clumped on all scales. Fig. 3, for example, illustrates the position of the brightest galaxies in the Northern Sky and shows their very noticeable clustering. It follows that a given galaxy will see more close companions in one direction than in another. Since galaxies attract one another gravitationally, that galaxy will experience a net acceleration in the direction of its closest neighbours and such accelerations will, in the course of time, give rise to random velocities which are a direct result of the uneven clustering.

Now it turns out that such random velocities depend not only upon the degree of clustering but upon the masses of the clusters and galaxies and hence upon Ω. This means, as Allan Sandage, Gustave Tammann and Eduardo Hardy first pointed out in 1972, that a measurement of the random velocities of galaxies, and of their clustering, should lead to a value for Ω. This brilliant idea is important, for the galaxies to be measured can be those *nearby* and therefore contemporary, and hence the estimate of Ω so obtained will be free of the unknown evolutionary effects which so bedevil the other tests. Accordingly, since 1972 the whole emphasis of cosmology has switched to the measurement and statistical analysis of galactic velocities in nearby regions of space. The one basic limitation to the technique is that it can provide only a lower limit for Ω, measuring as it does only the dynamical effects of the material *which is clustered like galaxies*. Any other components of the Universe which are not so clustered, for instance a diffuse intergalactic gas of very high temperature, if it exists (see Chapter 10), will not give rise to velocity disturbances amongst galaxies because it will pull them equally in all directions. Hence Ω (dyn), as we shall call such a dynamically determined Ω, is necessarily smaller than Ω (true). Nevertheless, Ω (dyn) is of great interest because most of the matter we *can* see is in the form of galaxies.

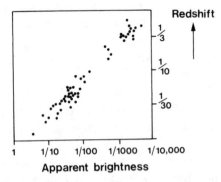

Fig. 13 The redshifts of bright elliptical galaxies plotted against their apparent brightness. If all the galaxies were identical and had zero random velocities, the points would lie on a perfect straight line. The rather small scatter about the line sets an upper bound on the random velocities of the galaxies, and hence on Ω (vis).

In their original paper Sandage, Tammann and Hardy analysed the Hubble diagram for giant ellipticals (see Fig. 13). They pointed to the very small scatter of points in the diagram about the mean line. The scatter should be large if individual galaxies have large random velocities of their own. They argued that hence Ω (dyn) must have a very low value close to 0. The original and unequivocal claim was so startling at the time that it provided a good deal of reassessment and controversy which will continue for some time. Sandage and Tammann have stuck to their original guess and in their latest paper boldly claim that Ω (dyn) = .04 ± .02, a value so low that it leaves no room for much material beyond the bare minimum we can already account for in galaxies. More skeptical astronomers analysing their own data have argued that the observations are consistent with any value of Ω (dyn) between 0.1 and 1.

To understand this surprising disagreement, we should look at the steps which lead to the computation of Ω (dyn). First, the astronomer must select a sample of galaxies whose velocities will be analysed. This is not trivial, for the chosen galaxies should be neither too near nor too far, and estimates have to be made as to the distance of each one of them. Secondly, the velocities must be measured accurately, which requires a great deal of telescope time. The random velocity of a given galaxy can be calculated as the difference between its measured velocity and its recession speed as estimated from its distance. The snag is that the recession speed, which can be calculated from its distance and the rate of expansion of the Universe, can be measured only to the accuracy with which the distance is known, that is to say, with an inaccuracy of plus or minus 50%. If one is not careful, the distance uncertainties will swamp the computation and the random velocities so measured will be completely spurious. Since the uncertainties rise at higher distances, only close-by galaxies, of which there are not so many, can be used. In fact, of the 82 galaxies in Sandage, Tammann and Hardy's original sample only the closest six or so are of any use for our purpose. To claim that the Universe is infinite on the basis of only half a dozen velocities is, to say the least of it, unconvincing.

But if one chooses instead a sample of closest neighbours where the distance uncertainties are not so worrisome, one runs into difficulties which are of two kinds. Firstly, the theory upon which the whole Ω (dyn) calculation is based breaks down for galaxies which are so close to one another as to be in gravitationally bound orbits. And secondly, although the relative random velocities of nearest neighbours may be small, they may nevertheless all be streaming at some high velocity relative to more distant galaxies, and this streaming velocity must be accounted in the evaluation of Ω (dyn). For instance, our local group could well be subject to the gravitational empire of the distant but massive Virgo supercluster (see Fig. 3), which will have imparted to us a net streaming velocity in that direction of perhaps several hundred kilometres a second.

Thus the observationalist is faced with a real problem in selecting a suitable sample which is neither too near nor too far, and where the optimum distance depends on the very Ω which he does not know. So the present dismaying uncertainty with claims and counterclaims for values anywhere between .05 and 1.0 is not so surprising as at first it seems. New television detectors have enormously speeded up the process of gathering galaxy redshifts and as the data set expands one must hope there will be gradual convergence towards a value for Ω (dyn) which is agreeable to all. The crux of the problem lies in getting more accurate distances for galaxies in order to subtract out properly their recession speeds.

The question of large-scale streaming motions is also the subject of much contemporary investigation and debate. The problem is to decide whether the nearby galaxies as a whole have a preferred direction of motion relative to the larger ensemble of galaxies surrounding them. The larger such streaming motions are the larger Ω (dyn) must be. Gerard de Vaucouleurs, who has made a life-long study of galaxies and who probably knows more about their overall properties than any man alive, has long argued that we are located at the outer periphery of a giant supercluster centred on the Virgo cluster some 10^8l.y. away. Observationally, the problem is identical to the earlier one, but here we look for correlations between the residual motions, that is to say differences between measured and recession velocities, and the positions of galaxies in the sky. For instance, if a local group is being sucked towards Virgo, then Virgo cluster galaxies will exhibit lower recession speeds for their distances than galaxies in the opposite direction.

If all the galaxies are wriggling and flowing around, it will be appreciated that a cardinal problem in all surveys is to locate a standard of rest against which such motions can be measured. However, following Penzias and Wilson's classic discovery the possibility was opened of measuring our speed relative to the cosmic background radiations (CBR). Whilst the theory of relativity denies the possibility of an absolute standard of rest, the CBR, which if it comes from the big bang, must be coupled to a large mass of material in the distant Universe, offers a close approximate to it. Our velocity relative to the CBR is therefore tantamount to a measure of our speed relative to the centre-of-gravity of the Universe. The CBR was, you will recall, visible light emitted by the hot dense gaseous material in the Universe at the time electrons and protons were recombining and space was becoming transparent for the first time. The photons we now see are redshifted from 3000 to 3 degrees only because that hot opaque curtain is receding away at a redshift of a thousand. They are, if you like, fossil messengers bringing information to us of the very earliest pre-galactic epoch. If only we could read that fossil record with sufficient precision much that is now a complete mystery might be revealed. Astronomers are therefore putting a huge effort into deciphering what surely must be the cosmic Rosetta Stone. They are looking principally for patterns or anisotropies in the radiation from different portions of space and at its precise colour or wavelength dependence. They know already that the curtain is uniform to 1 per cent or better, but surely at some level of attainable precision we shall see the writing on it showing through. In fact, we are just beginning to succeed.

Unfortunately, most of the CBR photons have wavelengths of around 1 mm which do not get through the atmosphere and which are in any case hard to detect. The seekers must cool their complex apparatus to within a degree or so of absolute zero with liquid helium and fly it above as much of the atmosphere as they can in flimsy balloons or cramped spy-planes like the U-2. In the meantime, theoreticians have been trying to predict what effects we might see. The motion of the galaxy and the rotation of the Universe should show up as slight variations of temperature on scales of large angular size. At a redshift of a thousand, regions more than 2 or 3 degrees apart were not causally connected and so corrugations on that scale are not unexpected. Nascent galaxies might show up as hot spots about a tenth of a degree in diameter, while the original radiation can be distorted or scattered by gravity waves or by intragalactic gas from much closer by. If the Universe has ever been reheated, as it might have been either by quasars or during the most vigorous epochs of galaxy formation, we can expect to see subtle colour changes in the spectrum. Predictions have been made as to the nature and size of all these effects on the radiation. Mostly they are very small, but their measurement is by no means beyond the grasp of conceivable technology.

American groups from Princeton and Berkeley have recently announced the first success. They agree in finding that one hemisphere of the Universe is hotter than the other by three millidegrees or about one part in a thousand. This so-called 'dipole anisotropy' is now firmly established and is most easily explained if we are moving towards the apparently hotter hemisphere at a speed of about one thousandth that of light or 300 km/sec. This motion apparently blueshifts or heats the photons in the direction we are going. Since the sun circulates about the galaxy at about this speed, the result seems not unreasonable. The pole of the hot hemisphere is not, however, in the direction of the solar galactic motion. It is nearly in the opposite direction. The inference must be that the galaxy and the local groups possess a peculiar motion of their own relative to the CBR amounting to about 500 km/sec in the direction of Hydra. Since this is 45° away from the supergalactic centre in Virgo we cannot attribute it straightforwardly to the pull of Virgo alone. But if the velocity is typical of galactic streaming speeds then it argues for a high Ω (dyn) in the range 0.3 to 0.7. Theoreticians are just now struggling to digest this exciting new piece of information and it may be some time before the full implications are worked out. We shall be happier when the temperature asymmetry, the galaxy streaming motions, and the local clustering can all be fitted into a coherent theoretical picture.

Evidence on Ω has also been looked for locally on the atomic level. The fact that hydrogen comprises 74 per cent of the visible cosmos by mass, helium 25 per cent and that all the remaining elements, including the oxygen, silicon, iron, carbon, nitrogen, so common in the terrestrial environment, make up only the remaining one per cent, presumably is telling us something about cosmic evolution on the atomic scale. Unravelling the relative abundances of the 100 stable atoms, and providing a cosmochemical rationale for them, has proved to be a fascinating and rewarding enterprise.

We now believe that only the three lightest elements, i.e. hydrogen (with its isotope deuterium), helium and lithium, were cooked up in the first few minutes of the big bang, while all the remainders have been baked piecemeal in the interiors of stars and thereafter dispersed into space by supernova explosions. It is, therefore, only to the light elements that we can look for evidence on cosmic matters like Ω.

In Chapter 6 we discussed how the relative proportion of helium to hydrogen probably came about in a big bang Universe and saw that the observed figure of 0.10 by number (0.254 by mass) is precisely what is expected when the 3° CBR is taken into account.

The density of the Universe (hence Ω) during the first three minutes naturally affects the rate at which some of the prevalent nuclear reactions can take place. The denser the plasma the more frequently nucleons collide and interact with one another, and so the relative proportions in which hydrogen, helium, deuterium and lithium have emerged from the big bang may yield information as to Ω. Calculations for various values of Ω_N lead to the results shown in Fig. 14. The subscript N reminds us that the density concerned refers only to nucleons and takes no account of more ghostly forms of mass as, for example, in neutrinos or gravitational waves.

We notice that whereas the helium abundance (expressed as a function, by mass, of hydrogen) is pretty insensible to Ω_N, the deuterium abundance falls by no less than ten powers of ten as Ω_N rises from 0.1 to 1. Deuterium should be a very sensitive probe of Ω_N, so what is its cosmic abundance?

Deuterium is a heavy isotope of hydrogen, that is to say it is merely

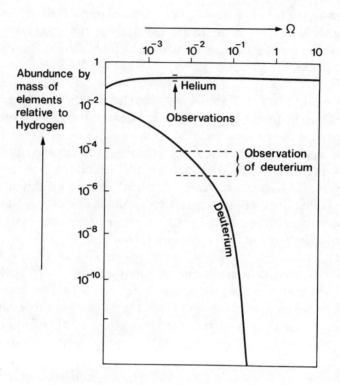

Fig. 14 Abundances of light elements as a function of the present density parameter Ω. According to the big bang theory the light elements helium and deuterium were formed in the early stages of the Universe by thermonuclear reactions in the primeval material. We see that the relative amount of helium is insensitive to the value of Ω, the cosmic density parameter. Deuterium, however, is highly sensitive and according to the calculations Ω could be no higher than a few times 10^{-2} to comply with the observation.

a hydrogen atom with an extra neutron stuck to the nucleus. Since the neutron possesses no electric charge it cannot affect the chemical properties of the atom, which means that deuterium and hydrogen are chemically identical. Wherever we find hydrogen we expect to find some fixed proportion of deuterium, and vice versa. With some effort, separation of the two isotopes can be achieved physically in the laboratory, for deuterium is twice the atomic weight of hydrogen. Incidentally, deuterium may one day prove the long-term energy future of mankind as the fuel for putative fusion reactors.

Where do we find large bodies of hydrogen? Locked up in H_2O, of course, otherwise known as water. Would it not be ironic to find, after all the searches we have made with giant telescopes and spacecraft, that Ω is most reliably written in a teaspoon of water . . . ?

> To see a world in a grain of sand
> And a heaven in a wild flower,
> Hold infinity in the palm of your hand
> And eternity in an hour.
> —*William Blake*

Seawater, or indeed any natural water, proves to have a deuterium abundance of 1.6×10^{-4} or 160 parts per million. Comparing this with our graph (Fig. 14) we read off a figure for Ω_N of about .01, leaving hardly enough room in the Universe for the galaxies we can already see. The Universe, it would appear, is open, and by a wide margin.

But there is a problem. The Earth can hardly provide a representative teaspoon of the cosmic composition when it is largely composed of elements like oxygen and silicon which are present in the wider Universe only in parts per million. If, using a telescope, we examine the bulkier atmosphere of Jupiter, the deuterium abundance there is lower by a factor of 8 at 20 parts per million. Even so, the implied Ω_N reaches only .03 or so.

The discrepancy between Jupiter and Earth suggests that solar system abundances are unreliable. They could have been affected by the still unknown processes which lead to the formation of our planetary system. We need to look further afield. The surfaces of stars can be examined spectroscopically but we cannot expect to find reliable deuterium abundances there because the substance is easily mixed down and burned in the hot interiors.

The cool diffuse interstellar gas between the stars is thought to be a mixture of unprocessed pre-stellar gas which has not yet condensed into stars and astrated gas, that is to say gas which was once part of a stellar interior but which has subsequently been dispersed when the parent stars exploded. Since astrated gas will have generally lost its deuterium in cooking, the deuterium abundance in the general interstellar medium should provide a *lower limit* to the primitive cosmic abundance, and hence, because of its slope in Fig. 14, an upper limit on Ω_N.

Interstellar atoms show up as dark absorption lines in the spectra of background stars. Unfortunately most of the interesting interstellar lines, including those of deuterium, occur at the ultraviolet wavelengths absorbed in the upper atmosphere. However, in 1970 Lymon Spitzer and his colleagues at Princeton launched a satellite telescope called Copernicus designed to explore the interstellar ultraviolent lines. Amongst a host of fascinating discoveries, they measured an interstellar abundance of between 36 and 4 parts per million along the lines of sight to several nearby stars. In view of astration, this points to an upper limit for Ω_N of about .02 and hence to a Universe that is open by a wide margin.

Some astrophysicists are impressed by this beautiful evidence, others point to the caveats. The doubters argue that with a trace element like deuterium, present only in parts per million, there might be other unsuspected mechanisms for producing such small quantities of any substance. Proponents retort that deuterium is more often destroyed in nuclear reactions and that in any case most of the proposed mechanisms are unworkable. But there are even more fundamental doubts. For instance, the calculations which lead to Fig. 14 rely upon a number of assumptions which may be reasonable but which cannot be checked. They assume there are very few neutrinos around (see Chapter 10) and that during the deuterium-producing stage of the big bang there was widespread isotropy. Since the causal horizon at that time encompassed a mere ten solar masses of material this is a drastic and presently uncheckable proposition. Any anisotropy would have generated extra deuterium and so lead us to infer a value of Ω_N much lower than the correct Ω. The cosmic lithium tells a similar story (Ω_N less than .08) but with parallel caveats.

The final global constraint on Ω comes about from chronological considerations. If Ω is large the expansion of the Universe is decelerating. If, therefore, expansion was faster during the past there must have been less time T_0 since the big bang. But T_0 must certainly exceed the ages of the Earth, the meteorites and the stars and this will set an upper limit upon Ω.

Radioactive dating of the oldest terrestrial granites yields an age for the earth of around 4.5 billion years, requiring Ω to be less than 10. The ages of the oldest stars are more problematical. Since all the stars in a single cluster probably formed at the same time, their age is inferred from comparing their observed properties with complex numerical calculations of their rates of evolution. For the oldest clusters ages in the range 10 to 20 billion years are deduced, arguing for an Ω less than about 3. While there may be considerable unseen matter, there cannot be limitless amounts of it.

Before summarizing the results of this chapter, it may be as well to point out that cosmologists seldom display a wholly dispassionate attitude towards the question of Ω. On the contrary, once their ideas are fixed they tend to become voluble advocates either of a closed or open

Universe. Partly this is in the nature of healthy scientific debate, but one cannot help feeling there is often more to it than that. Whether or not the Universe is infinite is a question with strong undertones of theology about it, even if the question can be settled on purely scientific terms. If the response is less than ecumenical, perhaps we should not be too surprised. At all events, in summarizing the evidence myself I feel it only fair to warn the reader of my own prejudices in the matter. Whilst they are not strong, they may still unconsciously obtrude into what I have to say next.

Personally I find it hard to believe that we have already identified all the predominant components in the Universe, a belief to which we must hold if indeed Ω is less than about 0.05. Considering the limited wavelength range within which astronomy has had to work throughout most of its history, considering the existence of entities like neutrinos and gravitational waves which are so ghostly as to be virtually undetectable, considering that observational astronomy has provided us with a perennial series of surprises to which there seems no end, and considering that in the nature of things we shall see only more and not less of the Universe as time goes by, I tend to opt for a high value for Ω around 1. Moreover, I am impressed by the numerical coincidence of the observed Ω to the value of Unity and so would question why God, or whoever arranges these things, might fix Ω so tantalizingly close to the critical value Unity and then leave the necessary little extra mass out. Finally, astronomy will be a great deal more fun if much of the inventory of the Universe is still undisclosed. Contrariwise if Ω is very low we have seen most of it already and only the details remain to be worked out. Whilst I recognize that the last two preferences are wholly unscientific, I admit to holding them nonetheless.

If the reader has found this chapter to be a tangled and confusing review, I can only say that it reflects the prevailing uncertainty as it exists amongst the experts. The subject is very much alive and if it seems to be staggering in several contradictory directions at once, such drunken progress in science often foreshadows the taking of a quite crucial and irreversible stride.

What we should *not* do now is reweigh all the evidence and try to pick a 'best' value for Ω. But to remind ourselves of the arguments, and as an aid for future reference, I have drawn up the summary Table 7.1. For compactness it employs the two symbols $>$ and $<$. Thus '$\Omega < 3$' means 'Ω is less than 3'; '$\Omega > .01$' means 'Ω is more than .01'; '$0.3 < \Omega < 0.7$' means 'Ω is more than 0.3 but less than 0.7.'

TABLE 7.1

Summary of arguments about cosmic density Ω

Argument	Inferred Ω	Remarks
1 Mass of visible galaxies	$\Omega > .01$	This must be a certain lower limit
2 Standard candle galaxies	$1 < \Omega < 3$	Assumes no evolution. Very uncertain. Evolution corrections would probably decrease Ω as given by as much as 1
3 Standard stick test	$0.4 < \Omega < 0.8$	Ostensibly independent of evolution, but this claim is not always accepted
4 Galaxy random velocities	$\Omega(\text{dyn}) < 0.2$	a $\Omega(\text{dyn})$ always refers only to material clumped like galaxies. Thus $\Omega(\text{dyn}) < \Omega$ b Lowest values quoted depend on very few data points and ignore large-scale streaming
5 Anisotropy of cosmic radiation	$0.3 < \Omega(\text{dyn}) < 0.7$	Anisotropy measurement looks secure but inferred Ω depends on limited map of local matter distribution
6 Cosmic deuterium abundance	$\Omega_n < .03$	a Ω_n refers only to nucleons (ordinary matter) ignoring neutrinos etc. Thus $\Omega_n < \Omega$

7	Cosmic lithium abundance	$\Omega_n < .08$	b Assumes small measured quantities of these substances could have been produced only in the Big Bang and by no other process c Employs sweeping assumptions about isotropy and neutrino-abundance in the early Universe
8	Ages of oldest stars	$\Omega < 3$	Relies on difficult calculations of stellar evolution
9	Age of oldest terrestrial rocks	$\Omega < 10$	Certain upper limit to Ω

Finally, I would like to draw up a short list of conclusions to which I think most astrophysicists would accede and which will act as a basis for further investigations:

(1) The evidence for evolution in the Universe is now compelling. This rules out a steady state Universe in favour of the big bang model.
(2) In the big bang Universe the density parameter Ω determines both the geometry of space and the dynamics of expansion. If Ω is more than 1 the Universe is finite and will recollapse. Otherwise it is infinite and ever-expanding.
(3) Ω certainly lies in the range between .03 and 10. This value, uncertain as it may be, is intriguingly close to the critical value of 1 considering it could have been very much larger or smaller.
(4) Many of the observations suggest there may be as much as a hundred times more mass in the Universe than we can account for among the visible galaxies. No observation conclusively rules out this very large preponderance of 'hidden' or 'missing' mass.

Let us now leave the global stage to search for the forms this presently invisible material may take and the sites where it may be hidden. In the next chapter we shall interrogate the galaxies themselves, for they have been giving some very surprising and contradictory evidence of late.

8

Weighing Galaxies

SINCE we cannot agree upon a density for the Universe overall, the alternative is to study the contents constituent by constituent. If we can separately estimate the densities present of visible galaxies, of gas, of neutrinos and so on, we can later add the densities together and come at least to a minimum value for Ω. Since the visible material of the cosmos appears to be almost entirely gathered into galaxies, we shall begin, in this chapter, to investigate the weights of individual galaxies. We shall find that even quite ordinary galaxies like our own contain significant amounts of invisible mass locked up in a presently mysterious form.

To measure the masses of objects like stars and galaxies which lie far beyond our physical grasp we must rely on the effects and laws of gravitation. The gravitational pull of a star increases in proportion to its mass, and decreases with the square of the distance away from it. And we can estimate the pull from the effect it has on moving objects nearby. For instance, the Earth moves in an approximate circle around the sun with a radius of 150 million kms. To keep the Earth in that circular orbit and prevent it flying off into space the sun must exert a calculable pull upon us. The calculated pull can then be compared with the feeble effect of one lead ball upon another, measured in the laboratory, and so the sun's mass can be compared with the measured mass of the balls. Likewise the Earth's mass can be computed from the pull the Earth must exert to keep the moon in orbit.

If stars had planets we could use exactly the same technique to calculate stellar masses. In practice, however, our present techniques do not allow us to distinguish planets, even at the distance of the nearest stars. Fortunately a good proportion of stars are multiple; they have other bright stars in orbit around them. Application of the law of gravitation to the observation of their orbital motions yields masses quite reliably. It turns out that there are clear relationships between the luminosities and colours of stars and their masses, so that stellar masses can now be inferred from observations even of isolated stars without companions.

Analogous techniques can be used to weigh whole galaxies. For instance, spiral galaxies are known to be rotating and a spectropic measurement of the rotation speed at a given radius yields a mass for that galaxy. It measures not the whole mass, but only the mass interior to the point of measurement. It is important to emphasize this distinction, for galaxies have a considerable extent in space, and if you want to measure the whole extended mass then spectroscopic measurements must be carried out to the very outermost regions where the light is fading away into the background sky. Considerable practical difficulty is then involved, for the available signal may be too weak to measure with a conventional telescope and spectrograph. Fortunately, giant new radio telescopes such as the 1000-foot dish in Puerto Rico, or the Westerbork radio array in Holland, can now detect radio whispers from regions far outside the visible extent of spiral galaxies, and with rather astonishing results, as we shall later see.

The gravitational influences of galaxies do not terminate at the boundaries. Giant galaxies like our own have satellite dwarf galaxies and globular star-clusters far out around them in space. Pairs of binary galaxies orbit one another. Indeed, galaxies are rarely isolated; they exist in groups and clusters whose members, numbering thousands in large clusters, all exist in a common gravitational empire. Just as a planet's velocity leads to a calculable mass for the sun, so the observed velocities of galaxies in clusters lead to a figure for the whole cluster mass, a figure which includes the mass of any invisible component lying in between the visible galaxies. We shall be looking at clusters in the next chapter.

In what units shall we measure astronomical weight? I could say the sun weighs 2,000,000,000,000,000,000,000,000,000 tons; but even if we can remember the number of noughts the figure does not convey much. Because luminosities are generally easier to measure than weights, a more significant measure of mass is the weight of an object per unit of luminous output. For instance, the sun weighs 2500 tons per kilowatt of heat and light. Judged in this measure, the sun is about a million times less efficient than a torch. The difference of course is that the sun will go on shining for billions of years. But the point I am trying to make is that if the sun is typical then an awful lot of mass has to be associated with each unit of starlight.

But is the sun typical? Let us look, for instance, at Sirius, the brightest star in the Northern winter sky. On close inspection, it proves to be a double star with one very bright component, Sirius A, the star we can see with the naked eye, and another component, Sirius B, which is a white dwarf 100,000 times fainter. The stars orbit around each other every 50 years, which allows a computation of the component masses in the usual way. Sirius A, which is 22 times as luminous intrinsically as the sun, is only twice as massive. It weighs therefore only 260 tons per kilowatt output, as compared with 2500 tons for the sun. We say it has, relative to the sun, a mass-to-light ratio, which we call Γ (gamma), equal to 260/2500 or about 0.1. Conversely, Sirius B, which weighs almost exactly the same as the sun, produces only 1/450th as much light (because its luminous surface is so small) and therefore has a mass-to-light ratio $\Gamma = 450$.

Henceforth in talking of the masses of stars and galaxies we shall work always in the mass-to-light unit Γ. Although Γ can range between the extremes of .001 for luminous giants, and 1000 for red or white dwarfs, the average value of Γ for a typical population of stars of mixed ages and masses is in the range 1 to 5, figures easy enough to remember. Γ is also a diagnostically useful quantity. For instance, if we look at a particular galaxy and find it to have an overall Γ of 2, say, then Γ is roughly what we would anticipate if the galaxy is largely composed of normal stars. But if Γ is higher than 10 we have good reason to expect that something funny is going on: that there is a lot of underluminous or invisible material about. Furthermore, if we know Γ for a galaxy and we know its luminosity—which is easy to measure—then we can straightway calculate its mass in tons if necessary, recalling that a Γ of 1 corresponds, as it does for the sun, to a mass-to-light ratio of 2500 tons per kilowatt.

We are now equipped to look at the masses of individual galaxies and we should naturally begin with our own. Recall that Milky Way is a typical giant spiral and that we reside far out near the edge of the disc. Interstellar smoke veils most of the Galaxy from sight so that mapping our Galaxy and its mass as a whole is paradoxically more difficult than it is for much more distant nebulae. There is, however, one compensating advantage. By assaying the contents of space nearby we can see what a typical galaxy consists of on the nitty-gritty scale. We can find out the masses and luminosities of its typical stars and estimate the gaseous content of its interstellar space.

What we would really like to do is concentrate on a smallish volume of space immediately round the sun, say within 15 light years, then identify and weigh up everything within it. Actually, this is much harder to do than it sounds, because of observational selection. Of all the stars in the sky, how do you pick out our few nearest neighbours? It is no good selecting the apparently bright ones. They are mostly luminous giants a very long way off. Our true neighbours may be, and generally are, very faint and insignificant fellows concealed in the richest star fields of the firmament.

The clue is to look for angular velocities transverse to the line of sight. If you look out of a moving train, nearby trees seem to be whipping by much faster than those on the horizon. If you compare the plates of a star-field, taken twenty years apart, some few stars will be seen to have moved minute amounts because both they and our solar system are typically moving at 20 km/sec randomly through space. Amongst these high-velocity stars will be our neighbours. But to find them all, the whole firmament must be photographed twice, twenty years apart, and all the corresponding pairs of plates must be compared. This mammoth and painstaking task has been largely carried out by a single individual, William Luyten, a Dutch-American astronomer at the University of Minnesota. Much of modern astrophysics, including most of what we shall be talking about in this chapter, rests directly or indirectly on Luyten's life work.

Luyten's work leads to the following picture of the stellar population in the solar neighbourhood (see Table 8.1). Within 15 light years of the sun, we can find 61 stars. Of these, only Sirius A, Altair, Procyon A and α Centauri A are intrinsically brighter than the sun. The great majority are insignificant red dwarfs each 1/500th as luminous as our own star. Ninety-seven per cent of all the light in the volume is produced by the four brightest stars; indeed, Sirius A provides more than half the light by itself.

But if the light is distributed so unequally the masses tell a very different story. The luminous stars are only slightly heavier than the sun, whereas the red dwarfs are, on average, about one third the solar weight. But being so numerous, the red dwarfs supply more than half the total mass. We can sum up by saying that the massive stars produce the light but the light stars produce the mass. Notice that the averaged out Γ, which is the figure we are really looking for, is about 2.

TABLE 8.1

Stellar masses and luminosities near the sun

Type of star	No. of stars	Total luminosity (solar luminosities)	Average mass	Total mass (solar masses)	Average Γ
Stars brighter than sun	4	50	1.75	7	.14
Red dwarfs	49	< 0.2	0.3	15	100
White dwarfs	4	< 0.2	.75	3	100
Sunlike stars	4	2	.75	3	1.5
Total	61	52	Total	28	52/28 ~ 2

In astrophysical terms, we understand Table 8.1 as follows. The total energy that a star can produce over its entire life is proportional to the amount of nuclear fuel available and hence is proportional to its mass. But the *rate* at which it radiates that energy varies much more dramatically, in fact as the fourth power of the mass. This is because massive stars generate high central temperatures which pump the heat out more rapidly. Massive stars therefore burn up more rapidly and become white dwarf embers. Conversely light stars burn very slowly as red dwarfs.

In a stellar population of mixed ages and mixed masses the massive stars are short-lived but produce most of the light so long as they live, while the lighter stars, which have high mass-to-light ratios and therefore contain a lot of the weight, may go on shining feebly for more than the age of the Universe. For such a mixed population, we expect an average Γ in the range 1 to 5, very like the figure 2 found for the solar neighbourhood.

For future reference it is worth inquiring how a population of stars could give rise to a very high Γ in the range 100 upwards. One way would be to manufacture only light stars. There would be no luminous giants and Γ would be typical of a red dwarf population, i.e. around 100. The alternative would be to produce a single burst of only massive stars, stars which would quickly burn up leaving only remnant white dwarfs with Γs of a hundred or more.

In our Galaxy, within the vicinity of the sun, it would appear that new stars have, since the Galaxy began, been forming steadily out of the interstellar gas, mostly in the mass range between 10 and 1/10th of a solar mass, with more dwarfs than giants. We believe very much more massive stars become unstable and explode, whilst putative stars less than a tenth the solar mass simply cannot burn and will remain invisible as the so-called 'black dwarfs.' How much mass is locked up in black dwarfs is very uncertain, but there are theoretical arguments and observations which tentatively suggest they are not very common.

To the mass of stars in the Galaxy we must now add an estimate for the weight of gas and dust drifting in interstellar space. This material is too cool to radiate significantly but it can be identified through the obvious absorption features it produces in the spectra of stars shining through it. Near the sun this interstellar gas, almost wholly hydrogen in composition, adds an extra 30% or so to the mass density in stars, raising Γ from 2 to 3. Although this gas is not itself luminous it is nevertheless important to Γ, for it is the reservoir out of which new young stars can form, stars which will significantly increase the luminosity per unit volume. Where there is no gas, as for instance in the nucleus or the halo of the Galaxy, we cannot expect to find luminous young stars like Sirius and we expect the mean Γ to rise from 3 to about 5 or 6.

Judged from our careful assay of space around the sun it would appear the Galaxy is made up of stars and gas with a mean Γ in the range 2 to 6. We must now look at the dynamical observations to see if this estimate works out on a large scale.

At the edge of the Galaxy where we live most of the material appears to be distributed in a comparatively thin disc. Jan Oort, the distinguished Dutch astronomer, has suggested an ingenious way of measuring the total mass per unit area in this disc, including the mass of any otherwise invisible constituents there might be. There are some stars in the disc with sufficient vertical velocities to carry them high above the disc. But when they get too far above or below the disc the gravitational influence of all the disc material will pull them back again. Then they will oscillate up and down in the gravitational field of the disc. Observation of the motions of these stars as a function of altitude will yield the total weight of the disc per unit area, and the weight will include the contribution of any invisible constituents like black dwarfs. In practice, the observations are not unambiguous. Some as-

tronomers have concluded that there could be as much invisible material as stars and gas in our locality. All we can definitely conclude at
present is that the invisible component measured this way, and in this
locality, is not overwhelming. Consequently, it could raise the local Γ
from 3 to 6 but not by much more.

The usual way to get at the mass of a rotating spiral like our own is
to look at the speeds of material orbiting around the galactic centre.
The orbital speed $V(R)$ α a radius R can then be related to the total
mass $M(R)$ inferior to R. In fact, a simple calculation shows that

$$V(R) \; \alpha \; \sqrt{\frac{M(R)}{R}} \qquad\qquad 8.1$$

If then we can measure $V(R)$ as a function of the radius R we can see
how the total galactic mass $M(R)$ increases outward.

The way that the rotation speed $V(R)$ depends on radius R is called
'the rotation curve' of a galaxy. Since rotation curves are crucial to
what we have to say next, such curves repay a little study.

Fig. 15 shows the rotation curve of a hypothetically simple model
galaxy, a model that can be interpreted with the aid of equation (8.1).
We assume that in the centre, out to the radius R, the model galaxy has
a spherical bulge of constant density. The mass of such a bulge of
radius R will therefore be simply proportional to the volume, i.e. be proportional to R^3. (The volume of a sphere of radius $R = 4/3 \; \pi \; R^3$, i.e. it rises
as R^3.) In equation (8.1) we therefore expect $V(R) \; \alpha \; \sqrt{R^3/R} \; \alpha \; R$. In
other words the rotation speed should rise proportionally with radius,
as it does in Fig. 15 (c), out to the radius R. Actually, there is such a
spherical bulge in the curves of most galaxies; there is one in our own;
and the rotation curves do indeed rise linearly near the centre. So far
so good.

Now consider the very outside of the model beyond R_2. In the model
we assume that virtually all the galactic mass lies within radius R_2 and
that there is merely a thin sprinkling of gas beyond R_2, just sufficient
to allow of a measurement of the local rotational velocities. So beyond
R_2 the galactic mass is virtually fixed and equal to M_0, say. It follows
from equation (8.1) that $V(R)$ is proportional to $\sqrt{M_0/R}$ beyond R_2. In
other words, the rotational velocities should *decrease* with R once we
are outside the significant mass. We show that decrease at the outside
of the model rotation curve in Fig. 15. Between R_1 and R_2 the curve
may take some complex shape (indicated by dots) depending on the
actual mass distribution. There is now only one further significant
point to note. If, beyond R_2, it is found that $V(R)$ remains a constant,

independent of R, then according to equation (8.1) there can be only one conclusion. $V(R) = const = \sqrt{M(R)/R}$ means that the mass $M(R)$ must be increasing in direct proportion to the radius.

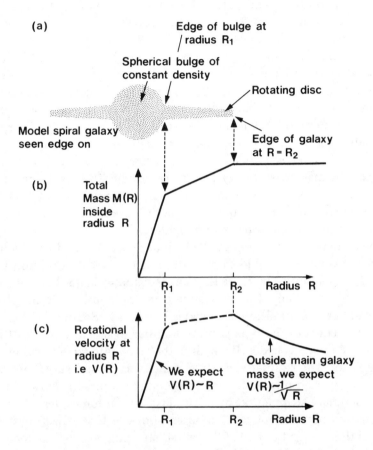

Fig. 15 The expected rotation curve for a spiral galaxy. (a) is a simple model of what such a galaxy looks like edge-on; (b) shows $M(R)$, the total mass inside radius R, as R increases. The edge of the galaxy is at radius R_2, and beyond R_2 $M(R)$ remains fixed. (c) shows the expected rotational behaviour and should be compared with what is observed (Fig. 16).

Measurements have been made of the rotation curve within our own Galaxy. The matter is not quite trivial because as everything is swirling around together it is not easy to find a standard of rest against which velocities can be measured. It is estimated nonetheless that the sun, which is 30,000 light years out, and very close to the galactic rim, is rotating around the centre at about 250 km/sec. Interior to the sun, the rotation curve is much what we expect if the mass and light of the Galaxy are in close correspondence. The inferred Γ for the interior material is about 6, a figure not so very different from the 2 to 3 in the solar neighbourhood and quite within the range expected of a galaxy made up of normal stars. This is all very satisfactory and postulates no high degree of invisible mass.

Out beyond the sun the galactic disc thins considerably, but there are some stars and some gas which can be used, in principle, to map the rotation curve much further out. For trivial geometric reasons, however, it is very difficult to distinguish between one type of rotation behaviour and another when you are inside the material. The surprising indications are, however, that the rotation curve remains flat for some way out. The implication must be that the Galaxy continues to increase in mass despite the lack of corresponding visible material. Our Galaxy may be surrounded by a considerable halo of invisible material. This is rather a dramatic conclusion and one we should not accept without strong confirmatory evidence.

The evidence comes from the behaviour of satellite bodies which lie out far beyond the visible extent of the disc, out half a million light years away or more. Out there lie weak irregular galaxies like the Large and Small Magellanic Clouds visible from the Southern Hemisphere, dwarf spheroidal or elliptical galaxies like that in Sculptor and globular star-clusters of the type that are familiar to us close by. All these mini and micro galaxies lie within the gravitational empire of our Milky Way. Their radial velocities, which can be measured, are controlled by it. And if they should occasionally venture too close-in their outer parts are stripped off by our Galaxy's tidal forces, leaving them in a noticeably sorry state. The observations of these dwarf systems, both of their velocities and of their tidal-stripping, leave little room for doubt. The mass of the Milky Way continues to rise far out beyond its visible extent. At the furthest measured point, which is twenty times further out than the sun is from the galactic centre, the mass is already ten times as large as it is interior to us, and it is still rising steadily in proportion to the radius out. The luminosity corresponding to this very massive halo around us is negligible, so it must be composed of material with a Γ in the range between 100 and 500. Adding this halo increases the total galactic mass by at least 10 whilst leaving the luminosity constant; hence the overall value of Γ for the Galaxy, including its halo, is about 10×6 or about 60.

If the Milky Way really has such a remarkable halo then it is probably not unique. What can we say of the masses of other spiral galaxies?

Clearly at optical wavelengths the rotation curve of a spiral can be measured only within that galaxy's visible extent. Indeed, until very recently, such observations of other spiral galaxies were confined by the paucity of light to the innermost and therefore brightest 30 per cent of the visible extent. The results so obtained were unsurprising, and led to Γs in the range 2 to 10, Γs typical of stellar populations like those near the sun, and close to the value Γ = 6 found for our own Galaxy inside its visible extent.

To search for dark haloes it is necessary to follow the rotation curves out beyond the visible light. This can be done in most spiral galaxies because they have thin peripheries of hydrogen gas which radiate in the radio region. However, even to 'see' these wraithlike peripheries requires a huge telescope tens of acres in extent; to measure a single such rotation curve requires a whole day's observations on a telescope a mile or more in extent. To reduce all those observations and turn them into a meaningful and reliable picture requires a man: a man with abnormal doggedness, persistence and integrity. In Groningen in Holland the right man and the right telescope were to be matched.

Albert Bosma, with his sky-blue eyes and thick blond beard, is very much a Dutchman. His sturdy build, ruddy face and slow-spoken voice bespeak more the farmer or fisherman than the scientist. Yet Albert spent the 1970s at the University of Groningen studying for his PhD in radio astronomy. He spent half a dozen years patiently collecting, collating and analysing many different radio observations that had been made of 25 nearby spiral galaxies with the Westerbork radio telescope which was just down the road. In 1978 he published a thesis entitled *The Distribution and Kinematics of Neutral Hydrogen in Spiral Galaxies of Various Morphological Types*. It was this unlikely vehicle which was to convince astronomers that there are significant amounts of invisible material in the neighbourhood of spiral galaxies, and most likely elsewhere in the Universe.

The really significant outcome of Bosma's work is illustrated by Fig. 16, taken directly from his thesis. It shows the radio rotation curves of 25 spiral galaxies measured out to radii which in most cases extend two to three times beyond the previous optical rotation curves and in many cases extend to beyond the last of the feeblest visible light. You can see that in nearly every case the rotation curve is approximately flat in the outer regions indicating an underlying mass that must be increasing in direct proportion to the radius. Although there are one or two discrepant cases (e.g. M 51), not one of them is consistent with the idea of a galaxy whose mass converges to a finite limit near the outermost visible margin. Indeed, the discrepant cases invariably belong to galaxies which are noticeably disturbed by the tidal effects of proximate companions.

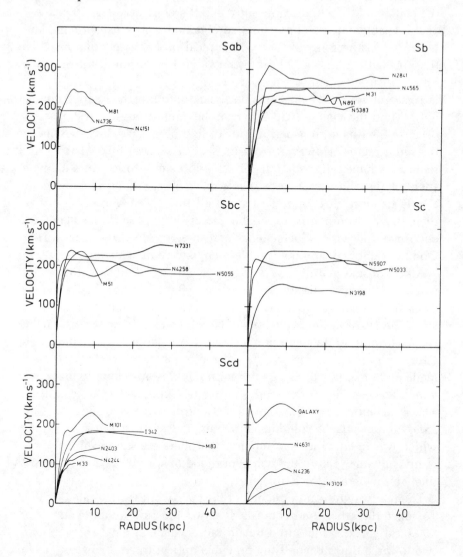

Fig. 16 Actual rotation curves of spiral galaxies measured by Albert Bosma with the Westerbork radio telescope. Each curve, labelled with the galaxy designation, e.g. N4151, shows the rotation speed of the galaxy in kilometres per second as a function of distance from the centre measured in units called kpcs, where 1 kpc is 3000 light years. The galaxies in each square are of a slightly different spiral shape. Notice the outer parts are flat, signifying hidden mass. Compare with Fig. 15.

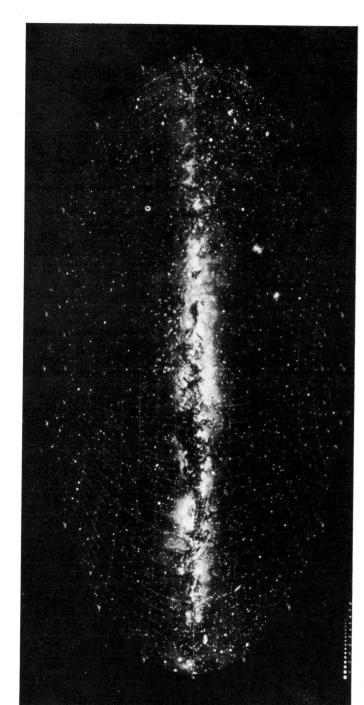

Plate 1 The celestial sphere: a photographic montage of the whole sky prepared from a large number of separate photographs and put together by the Lund Observatory. The Milky Way runs across the centre with the middle of the Galaxy in Sagittarius at the centre of the plate. That part of the Milky Way observable from the UK is the less spectacular region at the extreme right and extreme left. Notice the dark smoke lanes, and also the two Magellanic clouds in the Southern Hemisphere.

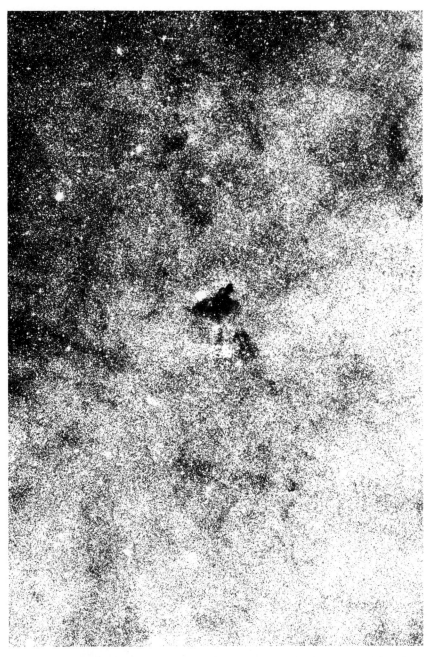

Plate 2 *Star clouds in the Southern Milky Way. The dark regions are not bereft of stars, they are simply obscured by clouds of foreground smoke. This smoke obscures all but the nearest regions in this photograph taken towards the centre of our Galaxy with the UK Schmidt Telescope.*

Plate 3 *The Andromeda galaxy. This nebula is the closest galaxy to us and is very similar to our own Galaxy. Note the dark obscuring clouds of smoke lying in the plane. Since we are situated in the outermost luminous parts of a similar galaxy the difficulty of discerning our own Galaxy's full structure through the smoke will be appreciated. The two fuzzy objects are dwarf elliptical galaxies, satellites of Andromeda, bound to it by its gravitational pull.*

Plate 4 *The globular cluster in Centaurus. This is one of the satellite star clusters which surround our Galaxy and which were used by Shapley to estimate its size. Each cluster contains about a million stars which buzz about inside it but which are prevented from escaping by their mutual gravitational pull. Elliptical galaxies are thought to be simply supergiant globular clusters containing up to a million times as many stars as this cluster.*

Plate 5 *A small cluster or group of galaxies in Indus. Note the different types of galaxy present, including a giant spiral in the centre with a pair of giant ellipticals to one side. Most of the fainter non-stellar objects on the plate are disc-like galaxies which belong to the group and do not have prominent spiral arms. The spiked objects are bright foreground stars in our own Galaxy.*

Plate 6 *A giant cluster of galaxies in Reticulum. Most of the objects on the plate, taken with the Anglo-Australian Telescope, are galaxies belonging to a single cluster containing thousands, and possibly tens of thousands, of galaxies. The three supergiant ellipticals, which are much brighter than our own Galaxy, may be cannibalizing their less luminous companions.*

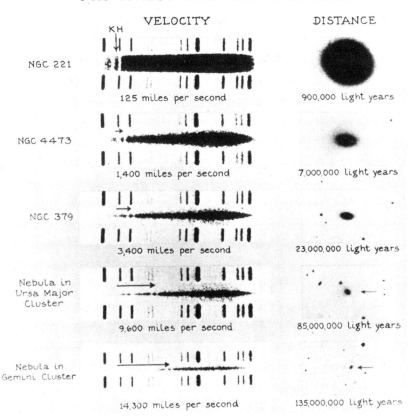

THE VELOCITY-DISTANCE RELATION FOR EXTRA-GALACTIC NEBULAE

	VELOCITY	DISTANCE
NGC 221	125 miles per second	900,000 light years
NGC 4473	1,400 miles per second	7,000,000 light years
NGC 379	3,400 miles per second	23,000,000 light years
Nebula in Ursa Major Cluster	9,600 miles per second	85,000,000 light years
Nebula in Gemini Cluster	14,300 miles per second	135,000,000 light years

Plate 7 *The evidence for an expanding Universe. On the left are the names of the galaxies or 'extra-galactic nebulae' as Hubble and others called them. The spectra (centre) show the Doppler velocities of recession and refer to the elliptical galaxies that are represented on the right with their distances from us, as estimated by Hubble, written in below. The photographs are all negatives. Note how the recession velocities rise systematically with the distances. Arrows on spectra mark absorption lines that are systematically moved towards the right (red) by increasing Doppler effect.*

Plate 8 The 'iceberg effect'. Two photographs of the same area of sky illustrate some of the difficulties of determining the relative brightness of galaxies. On the upper photograph the left-hand elliptical appears the brighter of the two, but on the lower and longer exposure plate the situation is reversed. Present techniques do not allow us to say for sure which of the two is brighter, and what is true for these two galaxies may be true for many others.

Bosma's evidence, which has been amply confirmed and in some cases anticipated by American and German astronomers using very different techniques, seems to be quite irrefutable. While the light distributions fade gently into the sky the masses appear to increase steadily towards the outermost points where measurements can be made. Whereas the luminosities converge the masses do not; the inference must be that even more invisible mass lies out beyond the present boundaries of detectability.

We should be careful, however, to draw a distinction between what can probably be inferred and what has definitely been measured. While it is true that the measured masses increase directly with radius, Bosma's observations extend only a little beyond the faintest detectable light and thus directly measure masses only twice as large as the conventional optical values. These observations imply a mean Γ for the spirals in the range from 8 to 12 compared to Γs of 3 to 6 from optical rotation curves. There can, however, be no question that invisible material (that is to say material with a Γ in the range 100 to 500) exists in significant amounts in the outer haloes of spiral galaxies. The obvious question is: 'How much further do the invisible haloes extend?' We have found in our own Galaxy, using satellite galaxies as probes, that the halo extends to radii ten times beyond the province of radio measurements like Bosma's and that Γ overall is as high as 60. We shall look at analogous evidence for the other spirals in Chapter 9.

Not all galaxies are spirals. Elliptical galaxies radiate a significant fraction of the light in the Universe and so we look at estimates of their masses next.

Elliptical galaxies are much rounder than the disc-like spirals and they do not appear to rotate much. Since they contain very little gas, young blue stars cannot form within them, so it is not surprising to find they are much redder than spirals. With their very different stellar contents, it would not be surprising to find that their Γ is different from the spiral value.

Since they do not support themselves against mutual gravitation by rotation, the stars in ellipticals must buzz about at very high speeds like angry bees in a swarm. Such speeds, or velocity dispersions as we call them, can be seen in the spectra of ellipticals as a general broadening of the stellar absorption lines due to the Doppler effect (see Chapter 3). The higher the mass of an elliptical or the smaller its radius, the stronger the gravitational fields will be. The stronger the field, the higher the velocity dispersion required to resist it. In other words, the velocity dispersion at a given radius in an elliptical is a function of the total mass interior to it, just as in a spiral galaxy the rotation speed is an accurate probe of the interior mass.

Large optical telescopes fitted with spectrographs and the latest and most sensitive television detectors can measure the velocity dispersions in ellipticals and hence provide evidence as to their masses. The results accord very well with what we already know of spirals. Within the luminous cores Γs are found to be in the range 5 to 10. As we observe outward from the core, the velocities remain constant instead of falling away, as they ought to do if the mass distributions and light distributions were matched. There must, therefore, be increasing proportions of sub-luminous or invisible material in the outer haloes of ellipticals, just as we find in spirals. While the evidence is not so watertight as is the spiral case, it is nevertheless rather convincing. To be more certain, we need rather larger optical telescopes than we have today.

One startling instance is known of an elliptical with a truly enormous invisible halo which may weigh as much as 100 times the visible contents. At the heart of the Virgo cluster, the nearest significant cluster of galaxies to us, there lies a powerful elliptical galaxy known as M87. For some years it has been known that M87 is a copious source of radio and X-ray emission. In 1978 Riccardo Giacconi and his American colleagues launched the first imaging X-ray telescope into space, with M87 as a prime target. The observations reveal an enormous spherical X-ray halo round M87, fifty times larger than the optical radius. X-ray spectra taken with the same telescope tell us the X-rays are being emitted by diffuse gas with a temperature of no less than 20 million degrees centigrade. To hold such a hot, high-pressure gas in place and prevent it expanding into space requires that M87 exert an enormous gravitational field. The source of that field must be a gigantic halo containing at least fifty times as much mass as we can see in the visible image.

M87 is certainly an unusual galaxy, and we cannot generalize to say *all* ellipticals have such massive haloes. Conversely, the fact that other ellipticals are not found with such spectacular X-ray emission does not mean their invisible haloes are any less massive. We detect the halo of M87 by chance because sufficient hot gas is there to reveal its presence. Hot gas in clusters tends to settle in their centres, but galaxies do not. A galaxy right at the centre of such a cluster, like M87, may accumulate enough cluster gas to signal its invisible halo through X-ray emission, whilst an elliptical further out may simply lack the gas to make the presence of its halo known. I emphasize that the radiating gas itself is not a significant receptacle of mass. It is that large amounts of invisible mass are required to hold the observed hot gas in situ.

We have decided that the visible spirals have mass-to-light ratio Γs around 10 while the visible parts of ellipticals also have Γs of ten. While both types of galaxy would appear to have considerably more massive, but invisible, haloes, an average value of Γ for the visible 'stuff' out of which they are constituted can be taken as 10. Now $\Gamma = 1$ corresponds to a mass-to-light ratio of 2500 tons/kilowatt of radiant output, so with visible galaxies, whose luminosity we know, we must associate a mass of about 25,000 tons per kilowatt. If we can estimate the average value of galactic light output per typical volume of the Universe we shall then be equipped to calculate the mean density of matter due to galaxies and hence arrive at $\Omega(\text{vis})$ for visible material.

What we need is an accurate census of all the galaxies in a representative volume of space. We need to know the number of spirals and their luminosities and the number of ellipticals. Likewise, we should include irregulars, whilst we should not forget all the dwarf galaxies of various kinds which could easily contribute, as dwarf stars do *within* galaxies, the major proportion of all the mass. The representative volume must be larger in size than a typical intercluster distance, otherwise our estimate will be heavily biased depending upon whether we are close to or well away from a big group. Given that 95 per cent of all galaxies are gathered into tight clusters and groups which occupy only 5% of space, this is obviously important. But if we choose a large enough volume there will be an awful lot of galaxies to work with and many will be too far off to observe accurately. Nor will the reader be surprised to find the insidious effects of observational selection creeping in once again. For instance, dwarf galaxies cannot be seen as easily as giants, so it would be very easy to underestimate their true significance.

For all its uncertainties the following picture is generally accepted. Ninety per cent of the light in the Universe is emitted by spiral galaxies, only ten per cent by ellipticals. Dwarf galaxies, that is to say galaxies less than one hundredth the luminosity of the Milky Way, are ten times as numerous as giants. But being on the average 100 times as faint, they do not contribute significantly to the total light. The brightest galaxies are about ten times as luminous as our own, the faintest are a millionth as bright. Averaged overall such a population density of galaxies will contribute only one kilowatt of light for each immense cube of space one billion miles along each side. This is not much and we are obviously exceptionally lucky to find ourselves so close along-side a hot and luminous furnace like the sun.

Taking the above estimate for the luminosity density, and multiply-ing it by the agreed mass-to-light ratio for galactic material of 25,000 tons per kilowatt (corresponding to $\Gamma = 10$), we find an Ω(vis) for visible galaxies of .01 or 1 per cent. This is the basis for our ear-lier assertion that visible matter fails to close the Universe by a wide margin.

But is the above estimate for the luminosity per unit volume accu-rate? There are good arguments for suggesting it is only a lower limit to the true value. For instance, there may have been some observational selection against dwarfs. Then again the luminosities we have assumed for galaxies are only those contained in the most obvious perimeters of light. True galaxies do not come to an abrupt end; they fade away gently into the starlit sky. The outermost and sometimes neglected regions may yet account for significant extra light. And have we got the clustering right? What about low-surface-brightness galaxies that cannot be distinguished from the sky, or dense point-like galaxies in-distinguishable from stars? Above all, can we really assume that most stars are members of galaxies? Perhaps there is a thinly spread but nevertheless dominating component of light from stars existing in the immensities of extragalactic space.

To be certain we have not seriously underestimated the extragalactic light, we would like to ascertain some safe upper limit to its permitted intensity. If the heavens are truly aglow with neglected galaxy light then, as distant galaxies are isotropically distributed, we expect to see the dim but even glow of their light all around the night sky. But astronomers who try to measure that glow find it no easy task because it is confused with much more powerful 'local' sources of light. The upper atmosphere, heated during the day, glows at night. The local stars, both bright and faint, are everywhere. Then there is zodiacal light, sunlight reflected back to us from fine dust lying in the plane of the solar system. Thousands of measurements of sky brightness at var-ious latitudes yield the following apparent breakdown:

TABLE 8.2

Total	100%
Zodiacal light	45%
Atmospheric airglow	33%
Local starlight	15%
Remainder unaccounted for	7%

The remainder should include local starlight scattered off dust in the interstellar space of our own Galaxy. Such scattered light should be recognizable because it ought to concentrate, along with the dust, in the plane of the Milky Way. The best estimates yield 5½% for the scattered light, leaving no more than 1½% as an upper limit to the extragalactic background light (EBL for short). Though it does not sound like much it is nevertheless seven times more than we have so far accounted for with the easily visible galaxies. The possibility, therefore, remains that we could multiply Ω(vis) by seven and raise it to around 0.07.

Estimating a small quantity like the EBL by subtracting very much larger and somewhat uncertain quantities (like the zodiacal light) away from one another can never be a very reliable procedure. Recently, however, a Finnish astronomer called Mattila has come forward with a very simple, ingenious and potentially more accurate scheme for measuring the EBL. Mattila points out that there are very dark absorbing smoke clouds in our Galaxy. Whereas the foreground light in the cloud direction should be the same both in front of and beside the cloud, the cloud will definitely cut out the EBL. A straightforward subtraction of the measured brightness from inside and outside the apparent cloud boundary should, therefore, yield the EBL. Mattila's own measurements yield an Ω(vis) as high as 0.3. However, American astronomers attempting to repeat Mattila's work found Ω(vis) less than 0.1. As yet, no one has traced the source of the discrepancy and more work clearly needs to be done. All we can safely conclude for now is that Ω(vis) is certainly more than .03 and probably less than 0.3.

Support for this massive halo hypothesis has come from a quite unexpected quarter. The beautiful spiral patterns found in many flat galaxies present astrophysicists with an intriguing theoretical challenge. The patterns are seen to spiral only once or at most twice around the galaxy. Yet the rotation speeds of the stars tell us that such galaxies must have rotated a hundred or more times during their lives. Were the spiral arms what they appear to be, that is to say necklaces of gas and young stars orbiting around together, then the arms should have coiled up a hundred times, not once or twice. We conclude that whereas the spiral patterns remain more or less fixed, the stars and gas stream through them, bunching together within the necklaces, expanding apart in between. How does such an improbable, yet widespread, dynamical pattern arise?

Dick Miller, a computer-wizard and astronomer at the University of Chicago, decided to try and find out by building model galaxies inside a computer. The models consisted of a hundred thousand or more mass-points circulating in a disc under their mutual gravitational attractions. Each mass-point represents a star in the model galaxy. Each 'star' is imbued with an initial velocity while its subsequent trajectory can be computed by calculating the net force of the other 100,000 'stars' upon it. Simple in principle, the program requires enormous computer horse-power and a great deal of ingenuity to make it work.

Far from producing the spiral arms they were hoping for, Miller and his colleagues were disconcerted to find their neat discs breaking up into lumps within a single rotation. They altered the masses, they altered the velocities, they altered the number of 'stars'; but no matter how they tried, their disc galaxies broke up into irregular lumps quite unlike the beautiful spiral discs we see all about us in space, and in some of which we live. Other writers who tried to improve on Miller's work succeeded only in confirming that Miller was right: rotating disc galaxies 'simply don't work.' (See Fig. 17 on page 155.)

Apparently the only way to rescue this ridiculous situation is to provide each model disc galaxy with a heavy spherical halo of mass which surrounds the disc. This halo should contain, within the circumference of the disc, at least two to three times the disc mass. The implication is that real spiral galaxies are surrounded by massive spherical haloes which may continue far out beyond the visible spiral arms and so constitute the great preponderance of mass. Since haloes are not in fact seen, they must be composed of invisible material. If they do exist, such haloes would also explain Bosma's flat rotation curves.

This satisfactory convergence of observation and theory has, during the past year or two, convinced even the most skeptical astronomers. To quote from a recent authoritative review: '. . . we think it likely that the discovery of invisible matter will endure as one of the major conclusions of modern astronomy.'

We can summarize the cosmological consequence of all this as follows. The visible parts of galaxies have a mass-to-light ratio Γ around 10. Multiplying the counts of visible galaxies by this ratio yields an $\Omega(\text{vis})$ of only .01. Extragalactic-background-light measurements, crude as they are, probably allow no more than ten times more light in the Universe than we can account for with the present galaxy census. To close the Universe with galaxies alone requires that they have mass-to-light ratios around 1000. We already know that both spirals and ellipticals contain significant amounts of invisible mass probably distributed about them in spherical haloes. How far these dark haloes extend

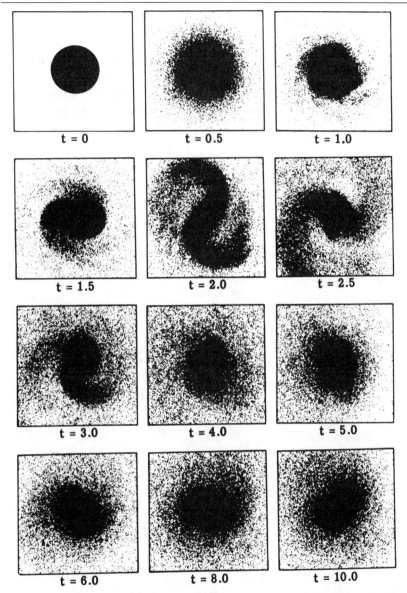

Fig. 17 A computer model of a spiral galaxy, consisting of 100,000 mass-points or 'stars' which start off (t = 0) rotating in a flat disc, though the individual 'stars' have in addition small random velocities of their own. Two rotation periods of the disc later, i.e. at t = 2.0, a sort of spiral pattern develops, mimicking the appearance of a real galaxy. However, the spiral pattern rapidly breaks up and by t = 4.0 the disc has broken into two lumps orbiting one another. This illustrates the instability of galaxies which consist of no more than their visible discs and hints at the presence of a great deal more invisible mass which is distributed spherically and which could stabilize the disc motions.

and how much mass they contain remains the question we must tackle in the next chapter.

It is worth remembering that our knowledge of masses on the astronomical scale is wholly derived from applying the law of gravitation, and applying it over enormous distances for which it has never been tested. If our law is not accurate, then all bets regarding mass are off.

Weighing Galaxies in Groups

$\mathbf{T}$HE suspicion that nebulae may have massive invisible haloes that extend far out beyond the last of the light and radio radiation is apparently confirmed for the Milky Way but needs to be tested observationally on a wide sample of other galaxies. While external galaxies do have faint satellites which can be used as probes of their gravitational power, alas they are mostly too faint for spectroscopic observations with the telescopes at our present command.

Fortunately, galaxies are seldom found in isolation. Five to ten per cent are found as binary pairs. Another ten per cent are to be found in great clusters containing hundreds, even thousands of members. The great majority are collected in smaller groups of from five to one hundred individuals. We believe these groupings are gravitationally bound together with the individuals typically ten to thirty optical diameters apart. Since we can observe the Doppler motions of galaxies within these groupings, we are in a position to measure the gravitational forces they exert upon one another, and hence to find their total masses, including any invisible components enclosed within the grouping as a whole. Even haloes which extend far out beyond the visible auras of galaxies are susceptible to detection this way.

Most of our direct knowledge of the masses of stars comes from studying binary stars. The obvious extension is to study binary galaxies. There is, however, a serious snag. Binary stars typically orbit one another over a period of fifty years. By the patient accumulation of observations the stellar companions can be watched in their elliptical progression and the complete parameters of their orbit, including the individual masses, worked out and confirmed. Binary galaxies, however, typically require a billion years for one progression of their stately dance. The fifty years during which man has been watching galaxies is far too short a time to spot even a hint of movement on a galactic scale. What we get is hardly more than a still photograph of a titanic pirouette. The actual motion, and the masses it implies, can be guessed at only by a lengthy and complex chain of inference.

If we tried to reconstruct the laws of soccer from a single still photograph of a single game, we should certainly fail. But if we had one hundred stills of one hundred different games we might arrive at somewhere near the truth. And that is the situation with binary galaxies. Apparently, close pairs are found in abundance in the sky. A fraction of these will be chance superpositions of a background and foreground object in the line of sight, optical doubles we call them, with no physical interconnection. The remainder will be true 'physical doubles,' that is to say, pairs of galaxies close-by in space and orbiting one another under the influence of their mutual gravities. To characterize completely a single such orbit requires seven separate pieces of information: these include the masses of the two components, the size and eccentricity of their elliptical paths and the orientation of their plane of orbit relative to sky. In practice, however, we can make only three separate observations of each double: the angular separation between the two galaxies on the sky, and the two line of sight velocities measured spectroscopically using the Doppler effect. With only three sevenths of the necessary information available, we need to resort to statistical arguments. For instance, if we had one hundred pairs with which to work, then, since we have no reason to expect that binaries will prefer any particular orientation in space, we can assume their orbits are randomly oriented with respect to the sky. Provided we have observations of enough pairs, then similar statistical arguments can be used to eliminate many of the individual uncertainties, allowing us to arrive hopefully at a reasonable value for average binary masses and hence of their mass-to-light ratios Γ.

The above preamble is necessary if the reader is to understand why the final results may still be the subject of dispute. He will also recognize that an observational study of binary galaxies is no light undertaking: a hundred radial velocities or more may imply hundreds of hours of delicate observation with the largest telescopes, either radio or optical. Above all, the observer wants to measure the velocity *difference* between the two binary galaxies, for it is the difference which largely determines the masses. For a typical binary that difference may be only 150 km/sec and to make observations with an accuracy much better than 150 km/sec is not at all easy, given the minute amounts of available light. Underestimating the errors or allowing for them incorrectly can lead to totally spurious results. The other pitfall concerns the inclusion of spurious optical pairs. They may be rejected because their velocity differences are considered to be too large. But if we reject all the high velocity pairs, we could well be biasing the results by rejecting all the most massive galaxies. So it is by no means simple to find an unambiguous value for the masses of galaxies in binaries.

The seminal paper on binary masses was published over twenty years ago by Thornton Page. From his pairs, pairs that are typically three diameters apart, and whose dynamics should therefore be sensitive to the presence of large massive haloes, Page found values of only $\Gamma = 1 \pm 1$ for spirals but about 40 $\pm$ 40 for ellipticals. While the rather impressive figure for ellipticals allowed for, but did not strongly support, the notion of invisible mass, the much tighter measurement for spirals definitely ruled such a notion out. For many years Page's study discouraged all enthusiasm for the really massive galaxy hypothesis.

Jerry Ostriker and Jim Peebles, two brilliant theoretical astrophysicists at Princeton University, New Jersey, had been much impressed by the breakdown of Dick Miller's computer models of galaxies. In 1973 they published a similar and very thorough computer analysis of their own. The analysis convinced them that without massive haloes it was impossible to stabilize spiral galaxies into the disc-like forms observed. This led them to re-appraise all the observational evidence as to the masses of galaxies, and to point out that Page's influential low mass results were entirely anomalous. Their concluding words, in a paper written in 1974, were: 'Further observational work on the puzzling problem of binary mass determinations would be especially rewarding at this time.'

The work was to come, probably not by coincidence, from Edwin Turner, a graduate student at the Hale Observatories. From 1974 to 1976 Turner observed 156 binary pairs and concluded that spiral galaxies possess haloes with ten times their visible disc masses, implying typical mass-to-light ratios Γ around 50, with ellipticals twice as heavy again.

This dramatic result stands in stark contradiction to Page's earlier conclusion. Why the difference and who is right? Although Turner was fortunate in being able to use more modern equipment, and in particular detectors sensitive to a wider colour-range of light, the largest part of the difference is probably more attributable to the presence these days of computers in every up-to-date observatory. Page was forced to do his extensive calculations with a pencil and paper. For sheer economy of effort he was led to work with the simplest of assumptions. For instance, he took all the orbits to be circular, which is strictly not true. Ideally, one likes to know how sensitive a conclusion is to any adopted simplifications. With a computer such an exercise is practical, for all the possibilities can be tried out in turn, with the results inter-compared. Working thus, Turner was able to argue that Page's statistical assumptions always lead to spuriously low estimates of the true masses.

Turner's very high masses for binaries are reinforced by Estonian astronomers and confirmed by work done with radio telescopes. The consensus view today is that spiral binaries are very massive with Γs of the order 30 to 60. As for the binary ellipticals, there are too few elliptical-elliptical pairs in Turner's sample to be certain they are not mostly chance associations along the line of sight. But the whole subject of binary mass determinations is by no means closed. Progress so far provides a sobering lesson to the dogmatist in his interpretation of any astronomical observations.

The vast majority of galaxies are found not in binaries but in loose groupings from five to a hundred individuals with typical neighbour to neighbour separations of about five to ten optical diameters. For instance, we belong to the local group with twenty to fifty members dominated by the giant Andromeda nebula and ourselves. The remaining family includes a much weaker spiral, several irregular galaxies like the two Magellanic Clouds which are consorts of our own, two dwarf elliptical companions of Andromeda and a ragbag of very low surface brightness dwarf objects of all kinds. There are no giant ellipticals. (See Table 4.1.)

If the galaxies within these groups are gravitationally bound to one another, then by a natural extension of the binary idea it should be possible to get at their masses. So the question is, are they so bound? After all, they could be expanding apart, or simply be chance aggregations of foreground and background objects upon the line of sight. What has convinced astronomers of late that they are bound is the sheer proportion of all galaxies in such groups. Recent studies indicate ninety per cent or more, far too high a fraction to be explained by chance superposition. If they were expanding we would have expected a goodly proportion to have dispersed already. So bound they would appear to be.

A way of looking at gravitational binding is to examine the balance between the random velocities of galaxies in a group and the forces holding them together. The higher the speeds the larger the forces and hence masses needed to prevent the group from breaking up. The velocities can be measured spectroscopically, so if we agree that the group is bound we can calculate the necessary binding forces and arrive at the implied masses and mass-to-light ratios. The uncertainties of the method exactly parallel those in the binary technique. Which galaxies are truly members of the group, and which are foreground or background objects accidentally in the line of sight? And how can we deduce the full dynamical situation when we possess only a single still picture and some of the velocities projected along the line of sight? And in some loosely bound groups the random velocities are 100 km/ sec or less, uncomfortably close to the observational errors.

To minimize the uncertainties, computers have been recruited in a number of cunning ways. Sverre Aarseth, a Norwegian computer-wizard, working at Cambridge, can generate groups or clusters of galaxies in a computer. He represents each galaxy by a mass-point, computes the net gravitational force upon each point of all the others, and then calculates the dynamical evolution of the whole ensemble. These models should accurately represent the behaviour of real clusters because all the true forces are included. The end product is a moving picture on a television screen of all the dots swimming about in space.

Now Aarseth plays the following game. Having added a reasonable number of foreground and background galaxies he freezes the picture and gives it to one of his astronomer friends. The astronomer now tries to estimate the mass of the group from exactly the same limited information he would observe from a real cluster. He makes his best estimate and hands it back to Aarseth. But in this case, Aarseth can play God because he knows the true velocity, position and mass of every 'galaxy' in the model group. In this comparison of the astronomer's estimates with the computer models, certain knowledge reveals the pitfalls and uncertainties of estimating true galaxy masses this way. The uncertainties are not negligible but it is hard to avoid the conclusion that galaxies in groups, that is to say the majority of all galaxies, have very high mass-to-light ratios in the range 40 to 80. These figures

accord with Turner's binary data and imply the presence of overwhelming amounts of invisible mass in the haloes and neighbourhoods
of most nebulae.

Chronologically, the first and still the most convincing evidence of this
invisible constituency, this silent majority of the Universe, comes to
us from a paper written nearly fifty years ago. The paper appeared in
the *Astrophysical Journal* for October 1937 under the title 'On the
Masses of Nebulae and of Clusters of Nebulae.' The author was Fritz
Zwicky.

If our story needs a hero, then Zwicky must fill the bill. Not only did
he write the seminal paper but he spent his life searching for new and
previously unsuspected forms of matter. But Zwicky was no conventional hero: he strode across the extragalactic stage like an irate colos-
sus, believing himself to be plagiarized and persecuted by some of his
most eminent colleagues. His life illustrates the personal passions that
lie, often concealed, beneath much of scientific research.

Though Swiss by nationality, Zwicky spent most of his working
career at the California Institute of Technology, the Institute associated
with Hale's famous Californian observatories. There for forty turbulent
years he fired off original ideas like a brilliant but angry firework.
Neutron stars, cosmic rays, gravitational lenses, supernovae, clustering, the missing mass, intergalactic matter and quasi-stellar galaxies all
attracted Zwicky's flashing intuition at various times in his career. He
even invented a new methodology for thinking which he entitled the
'morphological approach' and which he applied to jet engine design.
Not all his ideas are accepted today, even less found favour when
Zwicky preached them first. But if Nobel prizes had been awarded for
astronomy then, Zwicky might justly have won half a dozen in his
time.

While his illustrious colleagues, notably Hubble, Baade and Sandage, were using Hale's giant telescopes at Mount Wilson and Mount
Palomar, Zwicky was beavering away on a small 18-inch telescope of
novel design. His unique telescope, designed by Bernard Schmidt in
Germany, made up for its modest size with a field of view 1000 times
larger in size. The Schmidt telescope was ideal for surveying the Universe at large and Zwicky made excellent use of it. At the same time,
he felt himself to be unjustly excluded from use of the larger instruments. In the fiery introduction to his last book, an introduction he
headed 'A reminder to the High Priests of American Astronomy and to
their Sycophants,' he had this to say:

. . . the most renowned observational astronomers in the 1930s also made claims that now have been proved to be completely erroneous. This retarded real progress in astronomy by several decades, since the said observers had a monopoly on the use of the large reflectors of the Mount Wilson and Palomar Observatories, and inasmuch they kept out all dissenters. I myself was allowed the use of the 100-inch telescope only in 1948, after I was fifty years of age, and of the 200-inch telescope on Palomar Mountain only after I was 54 years old, although I had built and successfully operated the 18-inch Schmidt telescope in 1936, and had been Professor of Physics and Astrophysics at the California Institute of Technology since 1927 and 1942 respectively. E. P. Hubble, W. Baade and the sycophants among their young assistants, were thus in a position to doctor their observational data to hide their shortcomings and to make the majority of the astronomers accept and believe in some of their most prejudicial and erroneous presentations and interpretations of facts.

Whatever the truth of these accusations, and I have no idea whether Zwicky was largely justified or largely deluded, there is no question in my mind that Zwicky's 1937 paper is one of the greatest documents in astronomical history, to be counted alongside the works of giants like Copernicus and Kepler.

Zwicky begins his paper with a review of contemporary ideas on the masses of nebulae. He points out that rotation curves and estimates based on the supposed stellar contents of galaxies can yield only lower limits to their true masses. He goes on to argue, from photographs taken with his 18-inch Schmidt, that most galaxies are to be found in clusters. He then devises original techniques for getting at the masses of clusters as a whole. Then he applies his so-called 'Virial Technique' to the Coma Cluster, the nearest giant cluster to us in space, one in which he counted nearly 700 galaxies, the majority of them ellipticals. The requirement that the cluster be gravitationally bound against its internal velocity dispersion leads Zwicky to a mass-to-light ratio Γ for the cluster of no less than 500. This was a momentous and surprising discovery, for it suggested for the first time that in clusters at least, visible material comprised only about two per cent of all the mass. But Zwicky himself appears not to have been astounded. He believed that galaxies in clusters would tear each other to pieces during collision and that most of their substance would naturally be found as debris in intergalactic space. And on more philosophical grounds, he doubted anyway that men had more than a sketchy idea of the full population and richness of the Universe.

Zwicky's original work has since been repeated, extended and criticized many times over. Thousands of great clusters have been discovered with large new Schmidt telescopes and dozens have been weighed using Zwicky's technique. Criticism about foreground and background contamination and about large velocity errors are more easily refuted than in the case of small groups. The density of cluster galaxies on the sky is so much greater than it is around them that contamination cannot be too serious, particularly in the cores of clusters, whilst the measured velocity dispersions range up to 1000 km/sec as against expected measurement errors only one tenth as great. And computer techniques like Aarseth's can be used to investigate and eliminate some of the more obvious sources of error. The present consensus, based upon a wide variety of data, has it that great clusters have mass-to-light ratios in the range Γ = 200 to 400, very close to Zwicky's original value of 500 for Coma. Quite apart from the cosmological speculations of Chapter 7, Zwicky has forced us to believe that invisible material exists in significant amounts in the Universe.

We can now look back over the last two chapters and remind ourselves of the salient results.

In weighing the visible constituents of the Universe we have adopted as our unit of measurement the mass-to-light ratio Γ, where Γ = 1 corresponds to 2500 tons per kilowatt of light. Sunlike stars have Γs in the range 1 to 5. For short-lived giants Γ may fall briefly to .01 or less. Red and white dwarfs may have Γs up to 500. A mixed population of young and old stars is expected to have an averaged out Γ in the range 2 to 6.

If we count up all the obvious galaxies in a representative value of space, most of the light appears to come from giants: 90% from giant spirals and 10% from giant ellipticals. For such galaxies to close the Universe on their own (Ω = 1), they must have average mass-to-light ratios Γ of about 1000. Present measurements of the extragalactic background light of the sky allow for up to 30 times as much light coming from sources other than giant galaxies, but this is merely an upper limit not an incontestable measurement.

We have described many arguments for getting at the mass of nebulae. None is wholly satisfactory and all refer only to the mass as measured within a specified radius from the centre. Thus the weight derived from the optical rotation curve refers only to material enclosed within an area bright enough for spectroscopic observations; the radio rotation curve can probe the mass further out whilst information from a binary companion is sensitive to any material enclosed in the full extent of their mutual orbit. To each estimate of Γ we should therefore attach the maximum ranges measured in light years from the galaxy centre within which it applies. The results are roughly summarized in Table 9.1. The range of values quoted for Γ in each case partly reflects variations among individual galaxies, partly indicates the uncertainties of the method.

TABLE 9.1

Mass-to-light estimates for galaxies

Technique	Radius light years	Γ
1 Stellar population near the sun	30	2–3
2 Local stellar population ignoring young stars	30	5–6
3 Visible rotation curves of spiral galaxies	10,000	2–5
4 Radio rotation curves of spiral galaxies	30,000	5–10
5 Haloes required to stabilize spiral models	30,000	4–12
6 Ellipticals from internal velocity dispersions	3,000	5–12
7 Milky Way from satellite dwarf galaxies	600,000	40–70
8 Local group from timing argument	2 million	40–80
9 Binary spiral galaxies	1.5 million	40–80
10 Binary elliptical galaxies (tentative)	1.5 million	80–160
11 Small groups from velocity dispersions	1.5 million	40–90
12 Large clusters	2–6 million	400–600

We conclude:

(1) Spiral galaxies extend outwards to ten or twenty times their visible extent, with masses which appear to increase in direct proportion to the radius of measurement. Thus 90% of their mass lies in an outlying invisible halo.

(2) Likewise some elliptical galaxies have large haloes of dark material. Their mass-to-light ratios may be twice as high overall as among spirals.

(3) Ninety per cent of galaxies are in groups, largely composed of spirals. Γ for the groups, and for the constituent spirals, is around 50.

(4) Less than ten per cent of nebulae are in great clusters largely populated by ellipticals. Γ for the clusters is around 400 whereas the constituent ellipticals have Γs nearer 100.

(5) Quite apart from cosmological arguments the preponderance of material in the Universe is therefore in a dark, and totally mysterious, form.

(6) If we weigh up all the giant galaxies, even allowing an average Γ of 100 for each, the sum total of mass is only ten per cent of that required to close the Universe.

Finally, we should remind ourselves that galaxies and the hierarchies of groups and clusters into which they are gathered occupy a mere five to ten per cent of celestial space. Can we write off the remaining ninety-five per cent as a totally empty void?

Ghosts and Graveyards

TIME and again scientific discoveries have stretched man's credulity and imagination to the limit and beyond. The odd pauses when it seemed the dust was settling at last have all been upset by yet another earthquake. Our cardhouse theories collapse and we are left among the ruins to try and build again. Only a fool would predict that the earthquake season is over at last. For instance, it could easily be that we have overlooked the most important constituents of the Universe. They may be protected from our sight by the veil of the atmosphere. They may be too gigantic or too small for us to notice. They may pass through us like unperceived whispers. They may give off radiations we cannot detect or be hidden round corners of space and time. But if they exist at all, then, by definition, they must be perceptible to some sense and at some level of precise measurement. And as we shall see in this chapter man can be very diligent and ingenious in the pursuit and measurement of the most wraithlike products of his own imagination.

In the last chapter we argued that nebulae and clusters occupy only five per cent of all space. What lies in the gigantic voids between?

If the big bang theory is to be believed then galaxies have condensed out of a hot primeval gas. We do not at all understand how or when they formed but there is no good reason to believe that all the gas would have condensed, leaving nothing. If the formation of stars out of the interstellar medium is any model, then gravitational condensation is a grossly inefficient process which leaves up to 99 per cent of the gas in its uncondensed form. If indeed there is such a proportion of the original gas left over in intergalactic space then it would be just sufficient to close the Universe. Could we see this gas?

Since all elements heavier than lithium can be made only in stars then presumably the intergalactic gas consists almost entirely of hydrogen and helium. Once upon a time that gas was very hot and opaque because it gave rise to the cosmic background radiation which we can now see with radio telescopes. But since then the Universe has expanded, and so presumably has the gas. Expanding gas cools. The radiation has cooled from 3000 degrees to 3 degrees absolute, but expanding gas cools more rapidly than radiation and one calculates that the gas ought to be very cold indeed today with a temperature of less than one hundredth of a degree above absolute zero. Not only will it be cold but the gas will be rarefied far beyond the highest achievement of vacuum technology. Even sufficient to close the Universe amounts only to 1 atom per cubic metre. Since any given atom in such a cold diffuse gas will collide with a neighbour only once every hundred billion years, the radiation from the gas will be totally negligible. The only hope of seeing it at all is if the gas absorbs some of the radiation passing through it from a very distant hot object like a galaxy or quasar. Unfortunately, hydrogen and helium atoms are strong absorbers only of the sort of ultraviolet radiation which does not penetrate our atmosphere.

Were it not for a stroke of good fortune we might have remained ignorant of the state of the intergalactic medium to this day. But in 1965 the first quasar was discovered with a redshift of more than 2: 3C–9 it was called. The high redshift implies that we are seeing 3C–9 at a vast distance away, and therefore observing it through a long path length of intergalactic gas. More important, a redshift of two means that light originally emitted in the ultraviolet at 1200 Angstroms is redshifted to 1200 + (2 × 1200) = 3600 Angstroms which is in the *visible*. Now a cold hydrogen atom has its strongest resonance absorption at 1215A, so that if there was any absorbing hydrogen gas in the intergalactic medium it should show up as a prominent trough in the visible spectrum of 3C–9.

Amazingly, no such absorption trough was found. Nor has one been seen in any number of high redshift quasars discovered subsequently. We infer that intergalactic space contains less than one ten millionth part of the cold hydrogen gas needed to close the Universe.

The trouble with this result is that it is just too clearcut, too perfect. It implies that galaxy formation was a virtually perfect process which has mopped up all the gas leaving essentially nothing behind, and from what we already know of gravitational condensation this is almost impossible to believe.

The alternative is to suppose that the hydrogen, or at least some of it, is still there but that for some reason it cannot absorb. But the only way this could be so is if the hydrogen is so hot that it has become ionized, that is to say the electrons have all been knocked out of the atoms, thus precluding the 1215 Angstrom electron resonance. For the almost total ionization required to explain the observations, the gas temperature must be at least 10,000 degrees, and should probably be as high as a million. Such hot diffuse gas will radiate, albeit feebly, opening the possibility of direct detection.

The properties of the expected radiation will depend on the density, the clumpiness and the temperature of the gas. Since the radiation is produced by interparticle collisions the amount produced will rise sharply with the density, in fact with the density squared. A gas that is clumped into high-density clouds will be a much better radiator than a smoothly distributed medium, simply because more particle collisions will be going on per unit time in the denser clouds. The temperature determines what sort of radiation is produced; the higher the temperature the shorter the characteristic wavelength. The gas will radiate visible light at 10,000 degrees, soft X-rays at 100,000 degrees, harder X-rays at a million degrees. The temperature also fixes the pressure in the gas and if the pressure is too low the gas will simply condense into stars and galaxies under the force of its own self-gravitation.

We have already argued that any intergalactic gas left over from the big bang should be extremely cold now. But that argument assumed there has been no subsequent reheating of the gas by any of the dense luminous bodies, stars and galaxies for instance, which have latterly formed out of it. Quasars and nascent galaxies full of hot young stars were probably very potent blow-lamps, particularly in earlier times, so it is quite possible to contemplate an intergalactic medium that is as hot as 100 million degrees centigrade today.

We are principally interested to know whether an intergalactic gas can exist with the density $\Omega(\text{IGG}) \sim 1$ and be capable of closing the Universe, but without overstepping any of the observational constraints. To find out we put the density $\Omega(\text{IGG}) = 1$, assume a temperature, assume a clumpiness and calculate the radiant output.

A 'low' temperature (10,000°K) medium would emit principally in the visible where its emissions would be disguised from us by the luminosity of nearby galactic stars. But such a 'cool' medium is thought to be very unlikely because its low pressure would have allowed it to collapse into stars and galaxies. If a significant intergalactic medium still exists then it is likely to have a temperature of 1 million degrees or more and could be seen only by the glow of its X-ray emission. Has such an isotropic background of X-rays been found?

X-ray astronomy really began in 1962 when Riccardo Giacconi and his colleagues launched a small rocket in America. The USA was then engaged in its enormous effort to beat the Russians to the moon. Money for science was plentiful, provided research was confined to the immediate goal of exploring the solar system. To search for X-rays from the cosmos at large, as Giacconi wanted to do, did not fall into this category. But Giacconi was canny. He raised the money to search for X-rays knocked off the moon by solar radiations and flew a small detector for that purpose. If, during the course of its short flight, the detector happened to catch a glimpse of the sky elsewhere, who was going to worry?

From the point of view of its mission the flight was a total failure, with no X-rays from the moon. But from the point of view of astronomy as a whole the project was a momentous success. Very powerful discrete sources of X-rays were found in the constellations of Scorpio and Sagittarius and a general background of X-radiation was picked up emanating from all directions in space.

A new and exciting chapter in astronomy was begun: led by Giacconi researchers hurled a series of X-ray observatories into the sky. The Uhuru satellite of 1970 located 400 discrete sources many of which were tracked down (by optical astronomers) and identified with known celestial objects. Binary stars, neutron stars, quasars, active galaxies, supernovae, clusters of galaxies, even a black hole in the constellation of Cygnus were identified as copious sources of X-rays.

From our point of view X-ray astronomy has contributed three important discoveries. First of all there is indeed hot diffuse gas in the Universe. It is found within clusters of galaxies, has a temperature around ten million degrees and a mass of the same order as the visible galaxies in the cluster. It is called intra-cluster gas.

Secondly, the diffuse X-ray background detected on Giacconi's first flight is definitely confirmed. It is highly isotropic, suggesting that it comes from a very great distance away and it has a spectrum which is consistent with its origin in a ten million degree intergalactic gas of a density sufficient to close the Universe.

Thirdly, certain types of active galaxies and virtually all quasars are found to be copious emitters of X-radiation; in fact most of their energy is radiated that way. Such objects may possibly have reheated any intergalactic gas there might be, up to the requisite ten million degrees or more.

On the face of it the case for a significant intergalactic medium now looks strong. But there are difficulties. To begin with, X-ray astronomers at University College, London, have discovered that the intracluster gas contains normal (i.e. solar) amounts of iron. But iron can be manufactured only in stars and is not thought to be a product of the big bang. It therefore seems likely that the quite modest amount of gas observed between galaxies in clusters originates in the galactic stars and has been blown out into space by galactic supernova explosions. In other words it probably has nothing to do with the primeval gas we are seeking in true intergalactic space, and contributes nothing substantial to Ω.

In 1978 Giacconi launched his most ambitious instrument, an X-ray telescope called *Einstein* which provides true X-ray images of the Universe. *Einstein* confirms the earlier discovery made at the University of Leicester that there is a relatively common class of X-ray emitting galaxies. But it goes much further than that. If *Einstein* is pointed for a few hours in any random direction in space, a dozen or more faint X-ray galaxies and quasars are picked up per square degree. To the limit of detectability of the instrument there must be a hundred thousand or more X-ray point sources around the sky, and these must contribute about half the total X-ray background. The even larger satellite, AXAF, hopefully to be launched in the 1990s, will probably pick up even more and further sources, and it is quite on the cards that the X-ray background is simply the sum total of all those faint discrete sources. In other words while we cannot rule out a very hot intergalactic medium contributing significantly to Ω, the evidence for its existence is no longer compelling.

Gravity is a relentless force: the longer it acts the stronger it may become. Consider, as an instance, a tenuous cloud of gas, dust and smoke drifting about in interstellar space. The cloud is not entirely still for it is swept from time to time by hot gusts of radiation from evanescent young stars nearby, it is tugged this way and that by clusters, it may collide with another cloud, very rarely it may be racked by a supernova explosion in the vicinity. But drift as it may it is stable overall, for the weak forces of self-gravity are balanced by the gas pressure of its own atoms, heated to 50°K or so by the radiation field. But if, perchance, it should be significantly compressed by a collision, then an irreversible collapse may get under way when self-gravitation becomes dominant. The collapse is very slow at first and it is true that countervailing pressure forces will rise as the gas is compressed. But so long as the cloud remains transparent in the infrared, which it will do for a very long time, any compressive heat will simply radiate away leaving the pressure forces impotent to resist the mounting forces of self-gravitation. The more it collapses the stronger the gravitation becomes. The stronger the gravitation the faster the collapse accelerates. A vice-like, almost irreversible process is under way.

The collapsing cloud now has only three possibilities open to it. Should it become opaque then sufficient compressive heat will build up inside and raise the internal pressures into equilibrium with the gravitation. This is how a star condenses out of the interstellar medium. But it is very important to recognize that such a hot equilibrium, long though it may last, and it has lasted a billion years or more for the sun, cannot continue indefinitely. The interior heat supporting the equilibrium must leak away at last leaving the object once again at the mercy of gravity. Only if it can find a source of pressure totally independent of heat can it subside into equilibrium for eternity. Such sources of pressure are known: atomic pressures support the earth, electron degeneracy pressure supports white dwarfs, neutron degeneracy pressure supports neutron stars. However, we believe none of these 'cold pressures' can support against self-gravitation any object which is much greater than the sun in mass.

The second alternative for our collapsing cloud is for it to fragment into small pieces, each of which can separately support itself. Thus very large clouds may fragment into millions of stars, as we see in globular clusters, or into the billions which go to make up a whole galaxy. Once fragmented the pressure of the individual fragments rushing about, in other words the velocities of the swarming individual stars, resist the powers of self-gravitation, and the collapse finally halts.

But for a cloud either unable to stall off gravity by heating up, or unable to escape it by fragmenting, an inevitable crisis lies in wait. The collapse accelerates and accelerates until the cloud is halving in radius every year, every week, every hour, every second, every millisecond. . . .

Precisely what happens next must be a matter for speculation since no one has experienced or indeed observed such a catastrophe. But once the radius R has shrunk below the critical so-called Schwarz-schild radius R where

$$R_s = \frac{2GM}{c^2} \qquad \qquad 10.1$$

M being the Mass, G being the gravitational constant and c the speed of light, then the velocity required to escape from the gravitational field must exceed the speed of light. No light, indeed no radiation or signal of any kind, can any longer emerge. The object has vanished for ever from our ken and become a black hole.

As I said, the precise scenario is speculative. According to Einstein's theory, which pictures gravitation as a curvature of space-time, space has become so warped in the vicinity that it has closed in round the object cutting it off from the external world. It is as if we had placed a very massive dense ball on a rubber bed and seen the ball sink into the mattress so far that the rubber has closed in around and above it, leaving only a pucker on the surface to mark its disappearance. In other words, nothing remains beside the gravitational field itself.

Given sufficient compression, objects of any mass would become black holes. From formula (10.1) the Earth would become a black hole if compressed to less than 1 millimetre across, the sun if smaller than 1 km. The Schwarzschild radii, that is to say the critical radii below which objects fall into their own black holes, are calculated for various objects from equation (10.1) in Table 10.1.

TABLE 10.1

Critical radii for black holes

Object	Actual radius (cm) rough	Schwarzschild radius (cm) rough
Atoms	10^{-8}	10^{-52}
Man	10^2	10^{-32}
Mountain	10^6	10^{-10}
Moon	10^8	10^{-3}
Earth	10^9	10^{-1} (1 mm)
Sun	10^{11}	10^5 (1 km)
Globular cluster	10^{19}	10^{11} (3 light-secs)
Milky Way	10^{22}	10^{16} (10 light-days)
Cluster of galaxies	10^{24}	10^{20} (100 light-years)

The black hole concept stretches back to Newton and Laplace, but none has so far definitely been seen. You might suppose that by definition they were invisible, and that one could be lurking under the stairs. Provided the hole was relatively light, that is to say weighed no more than a mountain, then you could be right, because so weak is the force of gravitation that even at a metre's distance such a shrunken mountain would exert an imperceptible pull.

But massive black holes, in other words holes of astronomical weight, could manifest their presence in a number of indirect but potentially observable ways. A black hole of stellar mass might noticeably deflect the orbit of a binary star companion. Or if there were any gas nearby it would be sucked, or 'accreted' as we say, into the hole emitting a last scream of X-rays. A hole of galaxy-mass might deflect light-rays passing nearby and so produce a lensing effect upon the images of background objects. Gravitational tides could still be raised on neighbouring companions and Stephen Hawking, a brilliant physicist at Cambridge, has recently shown to everyone's consternation, and in spite of the enormous field, that black holes all evaporate, given sufficient time, releasing their energy by weird quantum effects, in the form of gamma-radiation. The required evaporation time though is inversely dependent on the mass: even a mountain-massed hole requires one entire age-span of the Universe to evaporate completely.

Bizarre as it may seem, the black hole concept is an insistent idea rather difficult to avoid. If one accepts three well-worn assumptions one is more or less forced by logic to accede to the idea. The first assumption is that gravity increases without limit as a body contracts, an assumption inherent in both Einstein's and Newton's theories. We know already of neutron stars with gravitational fields almost strong enough to blacken them out, and we know of no reason why such fields cannot amplify well beyond the limit. The second assumption, which is well tested by now, is that light has a constant finite velocity. The third assumption is that nothing can move faster than light. And physicists have spent fifty years trying to push particles beyond the light barrier in enormous accelerators with total lack of success. In theoretical terms, at least, the black hole is a respectable concept nowadays.

Astrophysicists have gone even further. They have worked out at least three scenarios whereby common entities within the Universe should become black holes as the logical end-point of their evolution.

Stars more than about 70 solar masses are never seen. If they did exist they would certainly be very luminous and easy to identify. Calculations show that such massive stars could be supported only by radiation pressure, which is a highly precarious form of support. Like a pencil balanced on its point a radiation balanced star can be upset by the slightest disturbance. If it is caused to expand ever so slightly the radiation pressure soon gets out of control blowing the star to smithereens; certain types of supernovae may be of this sort. Conversely a minuscule contraction would get out of hand in the opposite sense leading to a very rapid and unspectacular collapse into a black hole.

Even visible stars in the mass range 5–70 solar masses may finish in black hole tombs. Their stabilizing gas pressure can continue only so long as nuclear fuels can re-supply energy lost by radiation. But heavy stars burn up quickly and thereafter we know of no cold pressure that can save them from the grave. One can reasonably argue that the majority of stars heavier than the sun have either become black holes already or will become so in the course of time.

Whole star-clusters may form black holes of a thousand solar masses or more, by quite different dynamical processes. The stars which buzz around inside a dense globular cluster may nearly collide with one another from time to time. Should perchance three stars make a close approach then the probability is that one of them will be hurled out of the cluster by the combined gravitational fields of the other two.

Now the point is that the evaporated star carries with it, as the legacy of its own motion, energy which previously belonged to the cluster as a whole, energy which was needed to hold the cluster up against self-gravitation. In consequence, the residual cluster contracts a little and paradoxically the stellar velocities rise. A remorseless process is now under way. The more the cluster core contracts the more rapidly semi-collisions and evaporations take place. Since the cluster gravitational field must rise with each evaporation and each consequent cluster contraction, black hole conditions could finally come about in the cluster core. A thousand stars could be snuffed from sight at a glance. Thereafter the black hole could grow by gobbling up any further stars falling into its orbit. The whole process from start to finish requires no more than 100 million years or less than 1 per cent of the age of the cosmos.

In fact X-ray sources have been recently located in the cores of several dense globular clusters, and it has been suggested that we are seeing black holes of the above type engorging their prey. As the victim approaches it is torn apart by the tidal forces of the hole, releasing the X-rays from its interior before disappearing for ever from sight.

The above scenario, however, is by no means inevitable. Indeed when they tried to produce such a black hole in the Cambridge computer a rather different and surprising train of events took place. Binary stars were formed in the cluster following close encounters. These binaries beat up the other stars and chucked them out of the cluster. But the energy so lost was taken not from the cluster as a whole, but from the binaries concerned. Most of the stars evaporated all right, but they left behind not a massive black hole but one or more very tight binaries which, as we know, can produce X-rays in quite other ways.

But very dense, very massive star clusters of the sort which are known to exist in the nuclei of several nearby galaxies have no such binary escape-route open to them. The total gravitational binding energy, which rises with the square of the mass involved, is simply too large in nuclei to be packaged into binaries and evaporated that way. Regardless of the precise details it is difficult to see any way in which such nuclei can escape a black hole fate.

If you are of a skeptical turn of mind, as many scientists tend to be, then without clear incontrovertible evidence of the existence of a black hole in at least one case, you may prefer to remain unconvinced. But black holes, by their very nature, are elusive. Apart from Hawking-evaporation, which can be significant only in the case of mini-holes, a black hole will only give itself away by its action upon luminous material which is near to it. And as the Universe is a very empty place, the required close coincidence between a hole and a suitable visible tell-tale cannot be commonly expected. But if direct evidence is hard to come by at present it is worth looking at the circumstantial variety.

Either of the two fashionable theories of gravity leads us to conclude that when (see Equation 10.1) the mass of an object, divided by the radius R, exceeds $c^2/2G = 6.7 \times 10^{27}$ cgs units, then radiation can no longer normally escape from the surface, and the object is a black hole. The observer who has discerned an object and measured values for the mass and radius which satisfy the above criterion might reasonably claim to have 'seen' a black hole.

One of the very first X-ray sources to be discovered, and one of the most powerful, was found to be in the constellation of Cygnus and so was labelled Cygnus X-1. Since 1964 the source has been intensively studied from rockets, balloons and satellites, which reveal that the X-radiation crackles and flickers in a dramatic but irregular way. The flux may double in fractions of a second, may even vary on timescales of a millisecond or less. Since nothing can turn itself on and off in less than the light-travel-time across it, then Cygnus X-1 is probably no more than a light-millisecond across. By comparison, the sun is 3 light seconds in diameter, while even a white dwarf is one tenth of a light-second across. Cygnus X-1 must be a truly collapsed object; either a neutron star or a black hole.

All attempts to find a visible counterpart for the X-ray source failed until 1971 when radio astronomers located a weak radio source in the vicinity of the X-ray position. Since the radio output followed one of the longer term variations in the X-ray signal of Cygnus X-1, there could be little doubt that both signals had the same origin. Fortunately the radio observations pinned down the location rather exactly. Cygnus X-1 was found to coincide exactly with quite a bright star (only ten times too faint to be seen with the naked eye) already catalogued as 'Henry Draper 226868.'

Following this identification Paul Murdin and Louise Webster observed HD 226868 with an optical spectrograph at the Royal Greenwich Observatory. They discovered a blue supergiant of 23 solar masses at a distance of 6000 light-years. Although very hot and luminous by normal stellar standards, this star was far too cool and extended to be itself the source of the flickering X-rays. And anyway blue supergiants, which are not uncommon, are not usually the source of copious X-rays.

But Webster and Murdin made a more significant discovery. The velocity of HD 226868, as disclosed by the Doppler shift of its spectrum lines, varied from night to night, indicating that it was in orbit about an invisible binary companion, which itself has a weight in the range of 10 to 20 solar masses. Being invisible the companion is probably a collapsed object and the source of the X-rays. Dividing 10 solar masses by 1 light millisecond yields 6×10^{26} gm/cm, very close to the value required for a black hole. But the rationale for suspecting that the invisible companion in Cygnus X-1 is a black hole is much stronger than that. The rapid flickering argues for either a hole or a neutron star whereas the high measured mass precludes the latter possibility. Both observation and theory rule out neutron stars above 2 to 3 solar masses: cold neutron pressure is inadequate to support such weighty objects. So, by a process of elimination one is left with the strong suspicion that Cygnus X-1 contains a black hole. One supposes that the visible blue supergiant, like most stars of its type, is gently blowing off its surface layers by radiation pressure. When this stellar wind, as it is called, encounters the companion black hole it is sucked in and compressed down the throat emitting X-rays before it disappears. Because the wind supplies material unsteadily the X-ray emission flickers. It all fits together rather plausibly. Certainly no one has yet found a better explanation. However, if there were more than 2 stars in the system one might be able to wriggle out of the black hole conclusion. But not easily.

Black holes are also suspected to lurk in some galactic nuclei. About one per cent of spiral galaxy nuclei emit enormous amounts of X-rays and other high-energy radiation in an irregular manner. Variability and light-travel time arguments indicate radii for these nuclei of about one light-day. Unfortunately, the nuclear masses cannot be observed directly. However, lower limits to the masses can be inferred from the radiant outputs. Multiplying the observed luminosities by their suspected lifetime (at least 1% of the age of the Universe if we can see 1% of them active at any time) yields total energy outputs E for the nuclei where E, according to Einstein, cannot exceed Mc^2. When we divide the inferred mass of a 100 million suns by a radius of a light day we find numbers typical of a black hole. Providing such massive holes swallow no more than one solar mass of the dense surrounding material per year, either in the form of unhappy stars or of interstellar gas, they can provide all the radiation observed, radiation which can in some cases exceed by factors of a thousand the sum total of all the stellar luminosity from the surrounding galaxy. These nuclei, or quasars as they are called when seen at a great distance, are the most luminous objects in the Universe. Not bad going for 'black' holes.

Once we are agreed that black holes are, of their very nature, invisible, it seems to me we cannot turn up our noses at circumstantial evidence for their existence. After all, such witness is the most we can hope for. Their existence conceded, black holes then become the ideal explanation for missing light. Because mass is inferred from gravitational arguments, but not seen, all we have to do is conjure up a sufficiency of black holes: black holes in the halo, black holes in clusters of galaxies, black holes in intergalactic space. Like fairies black holes can be spirited up to explain almost anything. Unlike the holes in binaries or galactic nuclei, holes exiled to such dark immensities will always lack the fuel to swallow and so to give themselves away. They can be as black as you like.

Black holes probably exist. They can explain any mass-to-light ratio you care to mention. The posited prevalence of suitable numbers in the halo and so on cannot be refuted. The black hole hypothesis would appear to have everything and to explain everything. But if I were to countersuggest that all our mass deficiencies were owed to a population of coffins suitably strewn throughout space, my hypothesis would be likewise irrefutable. Coffins certainly exist. In sufficient (very small) densities they can explain any mass-to-light ratio we observe. Since they would quickly assume the temperature and therefore the colour of the 3-degree background they would be totally unobservable against it and therefore irrefutable. And why stop at coffins? Strawberries, corporation buses, abandoned space-ships, even dead fairies would do as well.

You will reject my coffin hypothesis, and rightly, not because it is refutable but because it is difficult to see how the coffins got there in the first place. But if coffins are improbable, what about black holes? Proponents of holes must provide a scenario for their emplacement which fits in naturally with whatever else we know about the Universe.

One thing we do know is that whenever a population of objects is formed in the cosmos there are usually more small ones than large ones: more sand-grains than boulders, more faint stars than bright ones, more dwarf galaxies than giants. The same general idea when applied to black holes, and to their precursors, provides some thankfully serious constraints.

The holes we have considered up to now are all 'secondary'; that is to say, they have probably evolved from previously visible material. For instance, the Cygnus hole was probably a massive supergiant star which ran out of fuel a million years ago. But for each massive star which condenses, we detect, and indeed find, many lighter ones which evolve more slowly. These lighter stars, which in *toto* should contain the preponderance of mass, should remain to be seen. But we have argued already that visible material cannot supply the majority of inferred mass, even in galaxies. It follows that secondary stellar holes like Cygnus should not contribute significantly to Ω.

From the rotation curves, nuclear holes in galaxies can supply only a per cent or two of the mass of the surrounding galaxy. They are insignificant on the cosmic scale. Indeed, for the arguments given above, we can probably ignore secondary holes of all kinds when considering the overall mass balance of the Universe.

Primary black holes, in other words holes formed as part of the primordial structure of the Universe, cannot be so lightly dismissed. For instance during the initial microseconds of the Universe the pressures and densities would have been enormously high, ideal conditions one might have thought for the production of black holes on all scales of mass from 10^{-5} gms up to galactic weights and beyond. Even in the big bang, though, we can find restraints.

The first is a theoretical restraint which arises from the very high temperatures which probably prevailed in the early cosmos. If we extrapolate the present 3° temperature of the cosmic radiation backward in time, we find that the early Universe must have been dominated by radiation pressure. For a black hole, or indeed for any structure to have formed, initial density fluctuations must have built up to the point where their self-gravitation could resist the general tendency to expand. But in a hot Universe calculations show that the radiation pressure would have torn such incipient structures to shreds. By the time radiation pressure subsided to a non-destructive level the primordial soup was already a tenuous medium unsuitable for the coagulation of black holes. This rather general argument explains why astrophysicists do not generally favour the generation of significant structures, be they holes, stars or galaxies, at very early times. To get round it we must suppose the big bang was either cold or tepid. But then one must ascribe the 3° background to some secondary cause, such as remnant radiation left over by a vanished generation of stars. Such theories cannot be entirely ruled out at present, but they are not favoured, on grounds of economy of hypothesis.

The other restraint on primordial black hole generation touches upon observations. Harking back to our arguments about the relative proportions of heavy and light structures in the Universe, we might expect the great majority of primordial holes to be very light, right down to the quantum limit around 10^{-5} gms. But Stephen Hawking has proved that light holes, i.e. holes lighter than a mountain, will all have evaporated already, creating as they go bursts of gamma rays. Gamma ray detectors flown above the atmosphere find no significant cosmic background, showing that truly primordial black holes contribute less than one hundred-millionth of the density required to close the Universe.

We have remarked already that very little is known of the way in which structures like galaxies actually formed. We do not even know whether galaxies condensed before stars or vice versa. Indeed, were we not confronted by the evidence on all sides we might, on purely theoretical grounds, be inclined to doubt that galaxies exist at all. What little we do know suggests that the most promising epoch for structural condensation occurred shortly after the Universe cleared, when the protons and electrons recombined. Recombination is believed to have taken place at a redshift of a thousand. Thereafter the radiation was decoupled from the matter and could no longer interfere with its condensation. At the temperature and densities which then prevailed, lumps down to the size of a million solar masses might have condensed. Lighter lumps would not condense because the gas pressure was still too great.

The collapsing pregalactic lumps might then have behaved in two ways. If they fragmented into stars, as calculations suggest they ought to have done, then globular clusters and dwarf galaxies would have resulted, and it has been suggested that these pregalactic clusters later agglomerated to form the galaxies we see today. But if, for some unspecified reason, the lumps could not fragment, then in short order they might have collapsed to become truly massive black holes. The gamma-ray difficulty we found for primordial holes is now obviated because such massive holes require far more than the age of the Universe to evaporate by the Hawking process.

We argued earlier that when a population of stars condenses, there tend to be more light ones than heavy ones. As we say, the initial mass spectrum is weighted towards the light end. But if the spectrum were to be changed we could have a very different Universe. Until we understand the origin of the present mass-spectrum we cannot be sure that in times past it was much the same. If an early, possibly pregalactic, generation of stars was formed with a mass spectrum weighted towards the heavy end, then the residue today would consist almost entirely of black holes in the size of 10 solar masses and upwards.

To summarize, we can say that black holes probably exist and they can in principle be used to explain the mass-to-light ratio Γ found in galaxies ($\Gamma = 10-60$) or in clusters of galaxies ($\Gamma = 300-800$). The problem is to explain how they came into being in the right mass range and in the right places. Primordial holes could not easily have formed in a hot early Universe, and if they did they should now be evaporating to produce a gamma-ray background that is not observed. Such holes as we have evidence for today, for instance in X-ray sources and galactic nuclei, are almost certainly secondary holes formed as the end products of evolution among previous luminous bodies like massive stars and star-clusters. As such, they cannot contribute significantly to Ω. The remaining possibility, and it is no more than that until we have some evidence to support it, is that holes formed at the time of galaxy condensation or shortly afterwards. These pregalactic holes would most probably exist in the range between 10 and a million solar masses. In the next chapter we shall see what observational constraints can be set on these bizarre and lightless bodies which, in terms of weight at least, could still dominate the Universe.

When a stone is thrown into a pond most of its energy is taken up by the ripples which spread outward. The energy cannot be said to reside in any particular droplets of water, for the droplets are stilled once the ripple passes by. So the energy, which cannot be destroyed, resides in the ripples themselves. Ripples, indeed any waves, are pure energy in motion. And the energy can be carried a long way. For instance, oceanic waves from Cape Horn can break on the coast of Cornwall.

Einstein's famous equation $E = mc^2$ has many consequences. Put in the form $E/c^2 = m$, it states that energy has a calculable, but generally very small mass. If, for example, we heat a saucepan of water we put energy into it, so increasing its mass by a corresponding amount (about one billionth of a gram per litre of water boiled). Now the Universe is full of energy in the form of waves and we should calculate their contribution to the mass density Ω. When we do so we find that star-light, radio waves, infrared, ultraviolet and X-radiation amount to a total $\Omega \sim .0001$, which is to say one three-hundredth of the contribution due to visible material. Most of this detectable wave density resides in the 3-degree cosmic background radiation.

Electromagnetic radiation, however, may not be the whole story. Imperceptible but nevertheless very energetic waves may yet rule the Universe. As an example of such a ghostlike wave, consider a tidal wave moving in the open ocean. Such a wave, triggered perhaps by an earthquake, could be 500 kilometres long and would be travelling at 500 kilometres per hour. If the wave were 3 metres high the energy content would be enormous: the equivalent of a hydrogen bomb. Nevertheless, such a wave would pass unnoticed beneath the keel of a ship because the long wavelength means that the whole sea-surface out to the horizon would rise and fall in unison in a period of one hour. The tidal wave is imperceptible because there is very little coupling between the wave and the passengers on board. Very little of its energy is absorbed by them. Similar tidal waves are raised in the earth's crust by the moon passing overhead. Your house rises and falls by a metre twice a day, but it does not fall down.

Now Einstein's theory explains gravitation as a warping of space-time and we have used the analogy of a rubber bed with heavy balls rolling about on it to explain the gravitational effects of massive bodies upon one another. If sufficiently violent motions exist then ripples and waves will be excited in the surface of the bed and these will carry energy away with them. Such are 'gravitational waves,' which are moving ripples in the structure of space-time itself. We would like to know what contribution they make to the mass-density of the Universe.

Gravitational waves are subtle entities, for if space contracts or bends so do the rulers embedded in it, and so do we, the observers. However, if we happened to be measuring the angle between two distant stars when a wave passed on by, then we would notice a transitory change in their angle until space had restored its normally almost flat condition. Moreover when a wave passes through a rigid ruler the ruler may go on shaking and ringing for some time afterwards. The shaking of a structure, not ascribable to any other cause, may be giving notice that a gravitational wave has recently passed by.

According to Einstein's theory the waves will travel at the speed of light and have periods characteristic of the motions which produce them. If you drop a brick it takes half a second to fall. The wavelength of the resulting gravitational wave will thus be a half light-second, or 150,000 km long. Now wave detectors are usually most sensitive to waves of roughly their own length. Ships, for instance, are rocked most violently by waves one ship-length long. You need a very large detector indeed to pick up the waves from a dropped brick. And most of the astronomical generators we can think of will emit even longer waves. Binary stars, for instance, generate waves a light-year long; binary galaxy waves will be billions of light-years in extent.

This practical difficulty may help to explain why gravity waves were neglected for so long. Although predicted by Einstein, they remained a mathematical obscurity for fifty years before Joseph Weber, a physics professor at the University of Maryland, came along.

Working almost alone Weber has rescued gravity waves from their obscurity and turned the whole subject into a respectable branch of physics. First, he reworked the theory into a more convincing form, then he sketched out various ingenious ways of detecting waves; finally he built an outlandish receiver to pick them up. Whether he has yet detected any remains the subject of debate.

The trouble with gravity waves from an experimental point of view is that they are very hard to generate, and even harder to detect. The output power P of a generator, which might for instance be a rotating bar-bell, is given by

$$P = \left(\frac{R_s}{L}\right)^2 \left(\frac{v}{c}\right)^6 \times 10^{42} \text{ kilowatts}$$

where L is the length of the bar, R_s its Schwarzschild radius (see equation 10.1 and Table 10.1) and v the velocity of rotation. From the formula we can see that only an apparatus that is compressed almost into its own black hole ($L = R_s$) and is rotating close to the speed of light ($v = c$) will generate much gravitational wave-energy. Even a normal pair of binary stars releases only a billionth part of its luminous output in the form of gravitational waves.

But this is a violent Universe we live in; Nature may have provided ideal conditions for generating gravity waves. When supernovae explode, when black holes collapse, when neutron stars collide, much of the energy released may find its way into waves. In such special conditions the expected periods are of order milliseconds and the wavelengths will be a mere 100 km or so, affording some chance that laboratory-sized equipment can pick them up.

Precisely because the waves are so hard to absorb they could go winging around the Universe for ever. Thus waves generated in a very early and possibly violent epoch of the Universe could be washing all around us, and through us, even today. Could the largest part of the Universe be locked up in the energy of wraithlike ripples left over from the beginning of time?

Following Weber's pioneering work efforts are now being made around the world to pin down this elusive energy. The waves should excite vibrations in the Earth, cause minute oscillations of the planetary system, set iron bars ringing, even interfere with the transponder signals travelling between spacecraft and Earth. But in all cases the expected signals are extremely weak, weak enough to be hidden by any number of supernumerary noise events. That is the challenge for the cunning experimentalist.

Of all the current attempts at detection only one has so far claimed any success. Weber himself set up a solid aluminum cylinder 2 metres long weighing 3½ tons. The cylinder, which is carefully isolated from seismic, acoustic and electromagnetic disturbances, has around its waist a belt of transducers. These pick up up any vibrations excited in the cylinder, and convert them into electronic signals. The cylinder is built to resonate naturally at about 1 kilohertz (i.e. one thousand times per second), a frequency close to the expected emissions from collapsing supernovae cores. With luck, the cylinder should trap a minute but detectable amount of kilohertz gravity-wave energy from any supernovae going off in our Galaxy. Such waves will be 300 km long, which means the pick-up efficiency of the apparatus is so small that the captured energy will scarcely exceed the random heat of individual atoms inside the aluminum. Like a gothic spire Weber's experiment bears witness to the faith, ingenuity and sheer cussedness of mankind.

Unfortunately even faith is not proof against all interference. Strokes of distant lightning, power surges and powerful cosmic-ray events high in the atmosphere above can all excite Weber's apparatus into spurious vibration. To eliminate these Weber set up another identical cylinder 600 miles away and operated the two together as a coincidence counter. Events detected in one cylinder, but not in the other, are discarded as local spuriosities. Those that remain, that is to say those which occur almost simultaneously in both cylinders, are then ascribed to gravitational waves.

In 1969 Weber began to detect one or more simultaneous events a day. Since his apparatus had some directional sensitivity, he was able to estimate where they were coming from. What convinced some of the onlookers was that the origin appeared to be fixed in relation to the stars, to be coming in fact from the Milky Way, and not to vary in concert with the diurnal or annual rhythms of the Earth and the sun. The same argument was used by Carl Jansky in 1936 to clinch his original discovery of radio waves from the cosmos.

Alas, euphoria has given way to skepticism. While Weber's observations cannot be faulted neither can they be repeated with similar or more sensitive apparatus. Furthermore, it was pointed out that Weber's positive detection, if it really emanated from the Milky Way, implied a generation rate so large that the Galaxy would be losing weight in a noticeable but unobserved way.

Today, we generally recognize Weber's experiment as a heroic failure. But like many such failures it has spurred others to redoubled effort. Extra funds, greater ingenuity, higher technology, are all being brought to bear with inevitable results. Every year sees the threshold of detectability pushed lower. If only a supernova would go off in the Galaxy most of the experimenters are confident they would pick it up. Unfortunately, the last one occurred 400 years ago.

Before we rule out a significant contribution to Ω from gravity waves, though, the detection threshold must be pushed down a factor of a further one hundred. For the present we can estimate $\Omega(G.W)$ only on theoretical grounds. Bernard Carr at Cambridge has recently surveyed the situation rather comprehensively and we shall look next at his conclusions.

As the Universe expands, the density of any wave-energy, be it gravitational or electromagnetic, falls away more rapidly than the density of matter. Conversely, wave-energy effects that are insignificant today could have overwhelmed the material when the Universe was highly compressed. That is the reason we believe the ordinary heat-radiation pressure, which is negligible today, probably tore apart any incipient condensations like protogalaxies, in the far distant past. Hence our difficulty in comprehending how galaxies formed.

Likewise, whatever the role of gravity waves today, much closer to the big bang they would have been far more significant. Now the greatest triumph of the big bang theory is its prediction that the Universe would have manufactured 25 per cent of helium by mass in its early stages. This prediction, however, which is borne out by all sorts of observations, is based on the assumption that even then the pressure of gravity waves was negligible. If, therefore, we set any store by the helium abundance as a support for our big bang theory, then we are forced to the conclusion that today $\Omega(G.W)$ is 10^{-4} or less. In other words, primordial gravitational waves, waves that have been rippling about since the beginning of time, are probably not significant in terms of their total mass.

Another argument concerns galaxy formation. Significant gravity-wave pressure would have made an already marginal process quite impossible, it seems.

A more significant constraint is set by the observed isotropy of the 3-degree background radiation. Waves present at the time of the recombination would leave one-degree-wide corrugations on the microwave sky that we do not see. Large local waves today should distort the microwave sky sufficiently to show up as a temperature anisotropy on the largest angular scales (90°), an anisotropy which is measured to be very small, if present at all. The observed lack of unevenness forces us to $\Omega(G.W's)$ of 10^{-6} or less if the waves are primordial.

If primordial waves can be ruled out as significant, what about secondary gravity waves produced by supernovae and so forth? Here it can be argued that violently moving bodies, be they visible or invisible, can only generate a mass of waves which is significantly less than their own mass. If we were to prove that black holes, for instance, had produced a mass of waves sufficient to close the Universe we should be forced to conclude that the black holes were ten times more massive still. In other words unless there is enough material to close the Universe anyway, secondary gravity waves will not help.

Until gravity waves are undisputedly detected, and until we measure the cosmic background radiation with much more precision from above the atmosphere, it seems sensible to discount the waves for now as major contributors to the universal mass density Ω. Whether or not they are significant on those terms they offer the exciting promise of probing the Universe to depths unimaginable before. While ordinary radiation is stopped by a few grams of matter, gravity waves could escape to us from the very jaws of a black hole, from the heart of a quasar nucleus, from the very instant of a supernova collapse deep in the core of a supergiant star. Gravity waves can even roll right across the Universe, effortlessly penetrating the opaque plasma of the early

Universe, carrying with them messages from the big bang itself. For these reasons gravity-wave research has only just begun.

There are reasons for believing that as you sit reading this book a billion billion tiny particles called neutrinos are passing clean through you each second. The energetic fraction of them have come straight from the furnace at the core of the sun, and of these only one will lodge in your body per year. The remaining and more numerous low-energy neutrinos have come much further. They were forged during the first second of the big bang and have been winging across space towards you ever since. So penetrating, so elusive are the primordial neutrinos that despite their numbers it is unlikely that one has ever lodged in the body of a human being. Individually these neutrinos are too ghostly and too light to weigh. Nevertheless, *in toto*, they might possibly contribute 99 per cent of the weight of the whole cosmos. They could possibly close the Universe on their own.

So unlikely is this claim, and so difficult to justify by direct experiment, that we must examine carefully the reasons why some quite respectable physicists believe it to be true.

One of the commonest modes of radioactivity is known as β-decay (beta decay). In essence what happens is that a neutron spontaneously decays into a proton and an electron (a β-particle, hence the name). Since the neutron mass is rather greater than the combined masses of the two daughter particles, the mass discrepancy can be transformed directly into energy via the $E = mc^2$ law. This energy appears in the velocity-energy with which the daughter particles are projected from the decay. Since the mass discrepancy is always the same the velocity-energy should likewise be invariant.

Careful experiments on β-decays made in the 1930s, however, showed nothing of the sort. The daughter particles emerged with all values of velocity-energy up to the maximum allowed by the mass discrepancy. For values less than the maximum, energy was simply disappearing. Careful measurements of β-decay showed further disturbing discrepancies. The momentum of the daughter particles was not adding up, as it should, to the original momentum of the parent neutron. And although electric charge was balancing out, spin-energy was not. This was all very worrying when you consider that the laws conserving mass-energy, momentum and spin are the foundation stones of any rational physics. The situation was only saved in 1929 by an imaginative idea from Wolfgang Pauli. Pauli postulated a new particle which was to be christened the 'neutrino' by Enrico Fermi.

This neutrino was to be chargeless and massless or very light, but it could carry off the discrepant energy and spin encountered in β-decays. Because of these peculiar properties, Fermi showed that the neutrinos would scarcely interact with ordinary matter and would therefore be virtually impossible to detect. An upper limit to the mass of the elusive neutrinos was set by the fact that on occasions most of the mass-discrepancy in a β-decay could be accounted for by the emergent velocity-energy. If neutrinos weighed anything at all they weighed 10,000 times less than the lightest particle previously known, the electron, and physicists generally assumed that they were massless, like photons of light-energy.

Once a new particle is hypothesized scientists normally go flat out to win a Nobel Prize by finding it. But for the neutrino the search looked hopeless. Fermi's calculations showed the interaction-strength or cross-section of a neutrino is so low that it can travel unimpeded through 50 light-years of lead.

This elusive intangibility is especially frustrating for the astrophysicist, for on his scale neutrinos could be significant in a number of ways. Neutrinos may bear off 5 per cent or more of the energy from the sun and since they can come straight to us from the nuclear furnaces at the very core of stars, potentially they bear information obtainable in no other way. Even more exciting, neutrinos could reach us carrying news of the first seconds of the big bang in which they formed. Such cosmic neutrinos would, however, be redshifted to low energy by the expansion of the Universe and Fermi showed that the lower its energy the harder a neutrino is to detect. In their frustration almost everyone forgot that light as they had to be, neutrinos might be present in such numbers as to represent in toto perhaps the dominant mass-component of the Universe.

The neutrino hypothesis, for it was no more than that, slumbered for thirty years. It was woken only by the advent of the nuclear reactor, for reactors should generate neutrinos in such numbers that detection of the odd one or two becomes a practical possibility.

In 1959 Clyde Cowan Jr and Frederick Reines assembled a huge scintillation counter outside the walls of the Savannah River nuclear reactor in America. Five hundred photo-electric eyes monitored a tank of twelve tons of water for the rare triple gamma-ray flashes that would signify the capture of a neutrino by a proton and its subsequent disintegration into a neutron and a positron as a result. With the reactor going full blast one or two such flashes per hour were indeed detected and Cowan and Reines duly won their Nobel Prize. No longer a hypothesis, the neutrino appeared to behave very much as Pauli and Fermi had guessed thirty years before.

The higher the energy of a neutrino the easier it is to detect. Nowadays neutrinos are routinely detected in high energy accelerators and it has been found that three sorts exist: the original electron neutrino, the muon neutrino and a new tau neutrino. Their properties are all very similar; all are very elusive, and in no case has it been possible to measure a mass.

Laboratory neutrinos were all very well but what astronomers wanted was a neutrino 'telescope' to detect the particles coming from outer space, and as a start from the sun. The whole theory of solar energy generation had been worked out in detail. Energetic neutrinos should be produced among the nuclear reactions going on in the sun's core. Now the theory needed a clear test.

The first 'neutrino telescope' has been built and is working several thousand feet underground in the Homestake gold mine in Lead, South Dakota. This unlikely telescope consists of 100,000 gallons of ordinary cleaning fluid in a large tank. The overburden of rock protects the telescope from all but the superpenetrating neutrinos. Once in a while an energetic neutrino from the sun's heart should hit a chlorine atom in the fluid, transforming it into a radioactive argon atom. Every fifty days the fluid is flushed out with helium gas which removes the argon. The argon atoms can then be counted one by one as they disintegrate radio-actively. Raymond Davies Jr, who runs the experiment, expected thus to detect about one and a half solar neutrinos per day.

The results have proved either disappointing or exciting, depending on your point of view. The good news is that Davies does detect energetic neutrinos. The bad news is that he is only finding about half a neutrino per day, or one third of the predicted number.

Hundreds, even thousands of papers have been written attempting to explain the discrepancy. Astrophysicists who have been blithely telling us for years how stars work are more than a little abashed when their sums turn out to be so wrong for the sun. None of the face-saving theories, which range from changing the sun's chemical composition to postulating that it is a variable star, enjoy much support at present and the whole issue is still very much open.

I would like to mention only one such theory, not so much because it is widely held but because it brings us back to the cosmologically significant question of the neutrino's mass. Elementary particles are known which continuously oscillate back and forth between quite distinctive states. It has been suggested that all three known types of neutrinos—the ordinary electron neutrino, the muon neutrino and the tau neutrino—are transitory states of a single particle which cycles round from state to state spontaneously. Now the Davies experiment is sensitive only to electron-neutrinos and detects one third of the number predicted to come from the sun. Could it be that the missing two thirds are accounted for by neutrinos which are indeed emitted but which are temporarily switched into the other two states? It is possible but cannot be confirmed as yet.

The cosmological significance of the neutron-oscillation hypothesis is as follows. A massless neutrino would necessarily have to move at the speed of light. But according to the theory of relativity, at the speed of light time slows down so much that, as seen from outside, nothing can change. Thus a massless neutrino should be incapable of oscillating between states. Conversely if neutrinos do indeed oscillate they must be moving at less than the speed of light and they must have a mass. Because of the sheer numbers involved even a very small mass might be sufficient to yield Ω(neutrinos) = 1.

This so-called 'massive neutrino hypothesis' has attracted a lot of attention of late. For instance, Lyobimov and others in Russia have remeasured β-decays very carefully and claim to observe a mass = 34 ± 14 eV, just sufficient to close the Universe. In an even more controversial experiment Reines and others in the States claim to see possible oscillations among neutrinos emerging from a reactor. Neither difficult experiment will be widely accepted without a great deal more confirmatory work and that is presently under way.

Disregarding their masses for now, we can still ask where the cosmic neutrinos came from, how many of them there should be, and where they fit into the wider cosmological picture.

According to the big bang model the Universe was very hot in its infancy. At such high temperatures protons, neutrons, γ-ray photons, electrons, anti-particles and neutrinos all freely transmute backward and forward into one another in a sort of statistical equilibrium and we should expect approximately equal numbers of each. After about 1 sec, however, the unreactive neutrinos decouple from the other particles and their numbers thereafter remain constant. For different reasons (see Chapter 6) the total number of photons is also conserved. Thus the number of cosmic neutrinos left in the Universe today should approximately equal the number of photons. Most of the other particles, like protons, should have been annihilated through colliding with their anti-particles, leaving only a small residue in existence today.

Now most of the photons about nowadays are to be found in the cosmic background radiation. The radiation temperature of 3 degrees implies about 300 photons per cubic centimetre distributed throughout all space. The equivalence of the neutrino-number and the photon-number, argued above, calls for about 300 to 1000 neutrinos for every cc of cosmic space today. Were each neutrino possessed of a mass no more than 30×10^{-33} gms, or 30 electron volts, then we should reach the closure density of 10^{-29} gms/cc.

The figure of 300 photons or neutrinos per cc is to be compared with the figure of 10^{-6} or less protons or electrons/cc, where the 10^{-6} follows from our earlier observation that Ω(vis) ~ 0.01. Photons and therefore neutrinos are, according to the big bang theory, 10^9 or one billion times more numerous in space, averaged overall, than ordinary particles. It is not difficult to see, if only they are possessed of a mass, why cosmic neutrinos could dominate the density and hence the geometry of the Universe. Paradoxically, though, these fundamental particles will be virtually undetectable. To catch a single high-energy solar neutrino per day requires a swimming pool of fluid. To catch one low-energy cosmic neutrino per day we would need a cubic mile of fluid, if the particles are massless. If they are massive enough to close the Universe all the fluid in all the oceans of the Earth still would not suffice.

How these ghostly particles will behave depends on their velocity, which in turn depends on their mass. If massless, they will hurtle at light-speed investing every nook and cranny of the Universe in equal measure, but affecting it in no significant way. If they are massive they could be whispering along at no more than 10 km/sec and could be exerting profound gravitational effects. Such slow 'massive' neutrinos would tend to clump into mass concentrations of their own or to fall into the gravitational wells produced by ordinary matter. We might find them dominating the mass in the nuclei and haloes of galaxies or controlling the dynamics of clusters of nebulae. Astronomers faced with the insistent problem of missing mass are now taking such a hypothesis very seriously indeed. If physicists can demonstrate for sure that neutrinos do oscillate and are therefore massive, then astronomy will be turned on its head yet again, with the largest, most luminous structures in the Universe dominated by the lightest and least perceptible. For the moment not all astrophysicists are equally enthusiastic about massive neutrinos and in the next chapter we shall return to discuss the main difficulties with the hypothesis.

In concluding this chapter I think it is worth remarking just how stimulating and how prescient theoretical arguments may occasionally turn out to be; how surprising and how far-flung the implications of an unexpected discovery may prove. The 1920s physicist probing at β-decay with his simple Geiger counter could never have imagined that his successors would be searching in gold-mines for the missing mass in the Universe, nor could Laplace, manipulating his equations, envisage his black holes belching out the enormous fire of a quasar. A hundred years ago it would have been unthinkable to suspect that neutrinos or gravity waves would be invoked to fill up the Universe. Surely the biggest surprises and the strangest discoveries are yet to come.

Taking Stock

$\mathbf{B}$Y now I hope that the reader is convinced that there are strong arguments for believing that the major, perhaps the overwhelming, proportion of all material in the Universe is presently invisible. What is this mysterious and invisible cosmic 'stuff' which makes up between 90 and 99 per cent of the Universe? Surely this must be the premier question for astrophysicists today. Many suggestions have been made which include dwarf stars, black holes, neutrinos, superhot gas, intergalactic stars and gravitational waves. In this chapter we shall weigh these and other suggestions and ask ourselves where, if anywhere among them, the balance of probability lies.

Before we start, it is worth assembling some of the general arguments that can be used for and against the various candidates. First we should look for direct independent evidence that the candidate actually exists: whereas dwarf stars are observed in millions, gravitational waves await their first detection, and until they are so detected the uncertainty must count against them. Secondly, we should look to see if the suggestion fits in naturally with what we already know of the Universe and with our currently preferred theories: for instance if the candidate is consistent with the observed element abundances so much the better. Thirdly, it should suit the rather detailed and unique observations we enjoy of our own Milky Way as it suits distant galaxies, for we find no reason to doubt that our Galaxy is a typical spiral. Fourthly, we shall, on grounds of economy of hypothesis, favour a candidate which can at one and the same time explain the mass discrepancies in individual galaxies, in clusters and in the Universe as a whole. Fifthly, we should naturally prefer a hypothesis which offers itself up to some practicable observational test.

Before beginning, we should define our terms rather carefully. The phrase 'the problem of the missing mass' has been used by various authors to describe two quite different problems. First there is the question of invisible mass in the haloes and environments of galaxies and clusters. There is no longer any question that the mass is there, because it shows up dynamically. What is missing is the light from that mass. Thus the problem should more properly be called the 'missing light' problem and we shall so refer to it henceforth. The phrase 'missing mass problem' will be reserved for the more hypothetical question of whether sufficient cosmic material exists, and in what form, to close the Universe. Until such time as we know $\Omega = 1$ we can still argue that this missing mass does not exist, whereas the missing light is a problem insistently forced upon us by the observations.

Our quest is therefore twofold. First, what is it that invisibly constitutes 90 per cent or more of the mass in and around galaxies and the groupings into which they are formed (the 'missing light' problem)? Secondly, *if* the Universe is closed, what extra constituents could make the cosmic density Ω up to 1 without conflicting with the observation?

Let us designate the mysterious agency which gives rise to the missing light by X. Spiral rotation curves tell us that X is spherically distributed around spiral galaxies in such a way that the total mass of X increases in proportion to the radius out from the galactic centre. A simple sum then tells us that the space density of X must fall off inversely as the square of the radius outward. The observations of binary and satellite galaxies tell us that these haloes of X continue out to between 10 and 20 times the visible radii, so the total mass of X in each galaxy is 10 or 20 times the mass of its visible stars. In order for X to remain out in the halo and not fall into the centre of the galaxy under gravitational forces, X must have a high speed in the region of two to three hundred kilometres a second. A comparison of the light and mass from these haloes tells us X must have a mass-to-light ratio $\Gamma(X)$ of at least 300 to 500.

For observational reasons the amount and location of X in elliptical galaxies is very much less certain. The lack of gas in ellipticals precludes the radio observation of long-range rotation curves, while the difficulty in isolating physical pairs of binary ellipticals in the dense clusters in which they mostly reside discourages binary mass determination. Nevertheless the optical measurements of rotations and velocity dispersions in ellipticals strongly suggest that X is present very much as it is in spirals, that is to say in haloes which increase their mass in direct proportion to the radius outward.

Including their haloes spirals have mass-to-light ratios Γ(spirals) not so very different from the Γs of the small groups in which they mostly reside (see Table 9.1). The evidence for significant intergalactic mass in groups is therefore not strong.

In great clusters, however, the mass-to-light ratio is ten times higher (Table 9.1) than has ever been measured in individual galaxies, even including their invisible haloes. Intergalactic X therefore probably makes up 90 per cent of the total mass in these clusters and in order not to fall into the cluster centres such intergalactic X must have a velocity around 1000 km/sec. Notice I am assuming that X is the same agency in spiral haloes, in elliptical haloes and in intergalactic space. This assumption is made to economize on hypotheses (Occam's Razor), though it must be admitted that there is no strong observational evidence to suggest that it is necessarily true. We return to this assumption later on.

So what is X the mysterious and invisible agency which gives rise to the problem of the missing light?

If you polled a sample of astronomers today as to the identity of X then the largest number would probably come out in favour of red and/ or black dwarf stars. Such dwarfs are the commonest stars in the solar neighbourhood and they can have a mass-to-light ratio Γ of 500 or greater. Many are ten thousand times less luminous intrinsically than the sun. Being so faint and red makes them difficult to detect nearby let alone far out in the galactic haloes and it would be natural to suggest that we have grossly underestimated their true contribution to the total galactic mass. Nevertheless there are significant observational and theoretical difficulties with red and black dwarfs as candidates for X.

Red dwarfs are stars with masses less than half but more than one tenth of the sun's. Calculations show and observations confirm that the luminosity of a star rises with the fourth power or so of its mass. Thus a one tenth solar mass dwarf would be an extremely feeble ember-coloured object emitting only one ten-thousandth of the radiation of the sun. If it possessed planets they would be frozen in perpetual night.

The internal temperature in such a dwarf would be only marginally sufficient to ignite thermonuclear reactions whilst the energy output would be so parsimonious that the dwarf could go on shining virtually for ever. Since the mass-to-light ratio Γ could go as high as 1000, the red dwarfs are obvious candidates for X. The question is, 'Are there enough red dwarfs?'

Being so faint, red dwarfs are hard or impossible to locate at any distance from us. But among the 61 closest stars (see Table 8.1) no less than 49 are red dwarfs and they make up 80 per cent of the total mass. This reflects the fact that when a family of astronomical bodies are born there tend to be more light ones than heavy ones; more asteroids for instance than planets, more interplanetary rocks than either. Observations of nascent stars in young stellar clusters show that stars of half a solar mass are born five times more frequently than sun-like stars, and so on. And as time goes on the disproportion between light and heavy stars becomes further exaggerated because the heavy stars burn up their fuel and go out. Hence the attraction of red dwarfs to those in search of X.

But there is a snag. Comparatively luminous stars like the sun have sufficient fuel to burn on for almost the expansion age of the Universe. Whereas, to judge from the neighbourhood, such solar stars do not contribute much mass in toto, they do provide a great deal, indeed almost all the light. Thus a small admixture of heavier sun-like stars amongst the dwarfs would reduce the average mass-to-light ratio sharply to about 5, far too small a value for the requirement on X.

Three ways have been suggested around this difficulty. The first is to suppose that the formation of heavier stars is suppressed in the halo or wherever the X predominates. Since no one has so far been able to suggest an astrophysical reason for this wholly arbitrary suppression I find this idea ad hoc and unattractive. It is fair to say, though, that nobody has been able to understand why light and heavy stars are born today in the proportions we observe and until this matter is cleared up the arbitrary suppression of heavy-star formation cannot be categorically ruled out.

A second way around the mass-to-light problem is the suggestion that we have missed a lot of red dwarfs; that red dwarfs of high Γ so dominate the population that even the admixture of luminous stars is inadequate to reduce Γ much below a hundred. But this idea is less plausible now that we have photographic plates sensitive to the sort of very red light that red dwarfs emit. Gerry Gilmore and Neil Reid from the Royal Observatory, Edinburgh, have recently completed a sensitive

infrared photographic survey of very red stars, a survey which singularly fails to turn up many more red dwarfs than earlier work using
blue-sensitive plates.

But what of stars which emit almost no light at all, black dwarfs as
they are called? Calculations show that objects lighter than one tenth
of a solar mass never generate interior temperatures sufficient to start
nuclear reactions at all. Such objects can shine feebly only so long as
they have heat energy left over from their process of birth. Black
dwarfs, difficult as they are to find, certainly exist, for a few have been
turned up more or less by accident. For instance van Biesebroek found
that of 600 nearby bright stars some 12 have very faint companions
which share the same motion in space as their brighter consorts. Some
of these are far too faint even to be red dwarfs and their masses are too
low (.05 solar masses or less) to sustain their own nuclear reactions.
We are seeing them only temporarily by the faint radiation left over
from their formative collapse. When this has leaked away they will
sink into dark oblivion.

Between the lightest red dwarf capable of nuclear burning (10^{-1} solar
masses) and the heaviest planet, Jupiter (10^{-3} solar masses), there is a
range of no less than 100 in mass. This gulf could hide a lot of black
dwarfs, a lot of invisible material. But it must hide it perfectly. Black
dwarfs while young (10^7 years) give themselves away radiating formative energy, and although they later become very feeble they do not go
out altogether. Indeed, the planet Jupiter is a very feeble black dwarf
still emitting, along with reflected sunlight, the last of its birthright
energy. Dwarfs in binary systems will perceptibly alter the velocities
and orbits of their more prominent consorts. So the black dwarf population in the neighbourhood, if not in the halo, can be roughly estimated from the observations. For instance, the Edinburgh red-star
survey turned up one 'black' dwarf only ten times more luminous than
the planet Jupiter. But while black dwarfs can be found they are not
numerous enough nearby to suggest themselves as the likely candidates for X.

I should be suspicious of this negative result, after all we are aware
of the pitfalls of observational selection, were it not for a buttressing
theoretical argument. Rather straightforward calculations show that it
is very hard for a collapsing gas cloud to fragment into independent
lumps much smaller than a hundredth of a solar mass, so it is difficult
to see how many black dwarfs could form.

Henceforth we shall lump red and black dwarfs together. This is
partly because their properties merge naturally into one another but

mainly because we expect them to form in the same way in the same places. Then if we can observe the longer-lived and more luminous red dwarfs we can make a rough allowance for the extra mass in black dwarfs which is probably no more than twice as great.

If there are dwarfs up in the halo of our own Galaxy they must have high speeds of more than 250 km/sec to get them up there against the galactic gravitational field. This compares with the random speed of 10 km/sec or so for stars like the sun which remain in the disc. Such high-speed dwarfs would not remain always in the halo: from time to time they should plunge down through our neighbourhood on their way round their orbits. Such close-by high-speed stars are easy to find; they will be among the rare objects which move perceptibly on our photographic plates from one year to the next. Maartins Schmidt and others have searched for halo red dwarfs among the fastest stars and they find only 1 per cent or less of the requisite numbers to provide for X. This is not very encouraging for the dwarf halo enthusiasts.

Strenuous efforts have been made to see the haloes around nearby spirals and weak haloes of a sort have been found. But they are neither red enough nor extensive enough to suggest an X of red dwarfs and their density falls off as the cube and not the square of the radius out.

Finally, an extensive halo made of dwarfs would be too 'sticky.' That is to say, if the galaxies with dwarf haloes approached one another, as they might easily do in binaries, in clusters and in groups, then they ought to stick to one another and merge into a single great lump as the dwarfs interacted gravitationally with one another. Computer calculations indicate that binaries and groups would quickly disappear in this way, whereas in reality they exist in healthy profusion. This is a strong argument against haloes made of dwarfs.

For the moment I find the evidence against dwarfs as principal characters in the missing light problem rather too strong. Their failure to show up in our own halo or in the haloes of nearby galaxies, the arbitrary requirement to suppress the birth of heavier stars in order to fix Γ at a suitably high value, and the stickiness problem, all argue against them.

Red and black dwarfs are congenital, that is to say they are born and remain frozen in their dwarf states. White dwarfs and neutron stars are evolved dwarfs, being the remnants of ordinary stars that were once very bright; that is to say they have sunk into their sad state through profligacy and overindulgence in early life. Many such evolved dwarfs are now known and as their mass-to-light ratios can reach 300 or more, they must also be considered as candidates for X.

Precisely how white dwarfs form is not known with certainty. But for them to have reached that state during the age of the Universe their precursors most certainly have been heavy stars, probably a few times more massive than the sun. With luminosities proportionate to the fourth power of their weight, such stars are comparatively short-lived. Since no star can be followed through its whole history, and because certain stages of our evolutionary calculations are still uncertain, we can only make an informed guess at the full train of events. Almost certainly the exhausted star will swell into an enormous bloated red giant. With its outer layers unstable it will then pulsate and, by one means or another, eject the greater part of its material into space. Only the very dense core, by now bereft of hydrogen fuel, will be left to cool off slowly as a white dwarf weighing around 1 solar mass.

Likewise, neutron stars are the tombstones of slightly more massive precursor stars probably around 10 solar masses. Such stars have a very violent death when the exhausted core suddenly collapses to nuclear density releasing enough energy to blow off the outer 90 per cent by mass in a cataclysmic supernova explosion. The neutron star remnant remains detectable for a few thousand years as a pulsar or X-ray source before it lapses into perpetual invisibility.

Probably only stars in the range of 2 to 10 solar masses eventually give rise to evolved dwarfs. But observations of present-day young clusters suggest that heavy stars in this range are born relatively infrequently: lighter stars in the red dwarf range are more commonly preferred. For instance, among the 60 nearest known stars (see Table 8.1) only 2 are in this mass-range and there are only 4 white dwarfs. Unless, for reasons we cannot fathom now, the balance of light and heavy stars was quite different in the past, then we cannot attribute the large halo mass to such evolved dwarfs. We should also remember that such stars have ejected the greater part of their substance into space as chemically processed gas rich in heavy elements like oxygen and iron. But the amount of such enriched gas in interstellar space is known to be rather limited. For this reason we do not see evolved dwarfs as major contributors to X.

Let us turn next to neutrinos as possible candidates for the dark mass. Three sorts of neutrinos are certainly known to exist; the real question is whether they weigh anything. A new and fashionable theory of elementary particles known as GUT (standing for Grand Unified Theory) allows for such a mass though it does not definitely call for it. Experimentally the neutrino mass must be very small, less than a ten thousandth that of the next lightest particle, the electron. But according to the big bang model neutrinos should be present in such large numbers that even light neutrinos could dominate the mass of the Universe.

At the present time three highly controversial experiments suggest that neutrinos may have some mass. Russian physicists claim to measure the mass directly; the American physicist Reines may have observed neutrino oscillations, oscillations which imply a non-zero mass, while the puzzle of the missing solar neutrinos would be neatly explained if such oscillations are taking place. But these results, and their implications, need independent confirmation which is still singularly lacking.

A chief attraction of neutrinos to the astrophysicist is that they can be used to jack up Ω without upsetting the good correspondence between the calculated and observed abundances of light elements like helium, deuterium and lithium. This correspondence, which is a chief attraction of the whole big bang hypothesis, breaks down if Ω is more than 0.1 and the mass is chiefly made up of ordinary particles like protons, neutrons and electrons. It does not matter whether these particles are now gathered in dwarfs, black holes or whatever. If the mass-to-light ratio $\Gamma \sim 500$ observed in big clusters is characteristic of the Universe as a whole (implying $\Omega = 0.5$), then rather more helium and considerably less deuterium should have been synthesized than we actually observe. But big clusters are rather rare and there is no guarantee that because $\Gamma \sim 500$ in them therefore Γ has the same high value elsewhere. But if it turns out that Ω really is larger than 0.1, then it may be difficult to rescue the big bang theory without resorting to neutrinos as the main receptacles of cosmic mass.

The snag with neutrinos is that their velocity is rather rigidly prescribed by physics once their mass is known. And given their velocity we can calculate exactly how they must behave and where they should reside.

The formula for calculating the neutrino velocity v is roughly

$$v = \frac{10}{\Omega_v}\text{km/sec}$$

where Ω_v is the total mass density calculated for neutrinos. If neutrinos served to close the Universe then $\Omega_v = 1$ (making a mass of 30 eV for each neutrino) and their velocity would be no more than 10 km/sec. Even the Earth orbits the sun at 30 km/sec while the sun itself moves around the Galaxy at 250 km/sec. Such low-speed neutrinos would tend to fall into the centres of galaxies and not remain in the haloes where they are needed to explain the missing light.

If on the other hand we set the neutrino velocity at the 300 km/sec required to stay in the halo, Ω_v would fall to 0.03 and the total of neutrino mass would scarcely be adequate to account for the dark mass required by the observations. And in big clusters the situation is much worse. There the neutrino velocity must be up to 1000 km/sec yielding a $\Omega_v = 0.01$ or a $\Gamma = 10$ which is far too low to be interesting.

This difficulty, compounded with the sheer lack of experimental evidence as to their mass, means the massive neutrino hypothesis has to remain something of a long-odds outside bet in my book, albeit an attractive one.

Neutrino detectors and neutrino experiments are everywhere improving. Cleaning fluid as a detector of solar neutrinos will be supplemented by a large block of rare gallium metal. Gallium picks up the low-energy neutrinos which evade cleaning fluid. What we would really like, though, is a big supernova to go off relatively nearby (but not too close, for safety's sake). If the escaping burst of neutrinos reached us at the same time as the light then the neutrinos, since they would have travelled at light-speed, would need to be massless. But if there is a delay the length of it will tell us the neutrino's mass.

Unless a black hole can feed on a rich source of nearby material then it should be virtually invisible. Black holes in the halo would find the grazing very thin indeed and so they suggest themselves as natural candidates for X.

We have argued already that nearly all single bodies above ten solar masses may collapse in rather short order to become black holes. Under present and neighbouring conditions massive stars like that are born only rarely, though it is conceivable that at an earlier epoch, or in a different environment, massive stars were the rule rather than the exception. In that case, their black hole remnants may comprise the major portion of cosmic mass.

Indeed there is some suggestion that a very early and very massive generation of stars should have existed in order to explain the present chemical composition of the Universe. This suggestion is prompted by the surprising discovery that the atmospheres of even the oldest-known stars in our Galaxy are not entirely bereft of heavier elements like iron and oxygen which must have been synthesized by an even earlier generation that is now invisible. To have disappeared by now, but to have left the correct legacy behind, it is calculated that the forerunner generation was composed entirely of massive stars. We should point out though that the iron and oxygen could have arrived by a quite different route. It may, for instance, have welled up from nuclear reactions in the stellar interiors of the stars concerned. Although calculations indicate that such upwelling is unlikely, we cannot be certain that such very old stars have not undergone shortlived periods of instability during their history.

Alternatively, there may be primordial holes which have existed since the very first beginnings of the Universe. The matter in such primordial holes would have been locked out from the early chemical synthesis. Then they could exist to provide a large gravitational field without overproducing chemical elements like deuterium. But the snag with such primordial holes, quite aside from their ad hoc nature, is that the lighter of them would already have evaporated by the Hawking process, leaving a residue of gamma-radiation that is simply not observed. Henceforth we shall ignore primordial holes and concentrate on the so-called pre-galactic variety.

We have discussed already our ignorance of the process by which galaxies and other structures developed in the early Universe. This ignorance leaves us a free hand to invoke black hole formation in any number of ways. It could, for instance, be that the majority of pre-galactic lumps collapsed to become very massive holes as heavy as galaxies or heavier. It is known, for example, that the epoch of recombination of electrons and protons, which first lead to the clearing of the Universe, was a specially favourable time for gravitational condensation to take place. Unfortunately, this happened at a redshift of a thousand and therefore at a distance which is only just becoming accessible to far-infrared satellite telescopes.

Tentative calculations suggest that most of the lumps forming at this epoch would have weighed as much as or more than a million suns.

What happened to these massive lumps . . . ? Again the calculations suggest they would have fragmented into clusters of a million or so ordinary stars. Very old globular clusters of about that size certainly exist (see Plate 4) and there are almost a hundred of them in our Galaxy, mostly in the halo. It has even been suggested that the galaxies we see today are rubbish heaps made up from the merging of millions of such globular clusters which have collided with one another and broken up.

Alternatively, the majority of such lumps may have collapsed without fragmenting into stars. Lacking any sufficient internal pressure they would then have sunk straight into black hole graves, emitting almost no light in the process. Such holes might go to make up galactic haloes today. What observational constraints can be set on an extensive population of such black holes with masses between about 100 suns and a whole galaxy?

A black hole whistling through the halo of our Galaxy will swallow any gas it encounters and spit out X-rays. The bigger the hole the more powerful an X-ray source it will appear to be. Limits on the number of X-ray sources observable in our halo suggest that any holes up there must each weigh less than a million suns.

If a massive hole charges through a star-cluster it will, by the force of its gravitation, tear out a few stars as it goes by. If there are lots of holes about, enough to make up the missing light, for instance, they will nibble away at clusters until there is nothing left. The fact that we can see quite healthy and very ancient clusters, both in the halo and in the galactic plane, limits the weight of such holes to less than 10^5 suns each.

In their encounters with ordinary stars the black holes do not escape entirely unscathed. Each such encounter will slow the hole down by a small amount. And as it slows it will gradually sink towards the galactic centre, having lost the energy it needs to climb back into the halo. The fact that the inner region of our Galaxy does not appear to be dominated by such exhausted holes may argue that each such hole has to be less than a thousand suns in weight. This would be a strong constraint if we could rely more securely on the tricky calculation which leads to it.

Finally, there is the stickiness argument that was used against dwarfs. You will recall that haloes made up of dwarfs would be so sticky that in groups and binary systems the galaxies would quickly merge into heaps, which is patently not the case. The same argument, but with greater force, applies to haloes made out of holes heavier than stars.

If we move out of haloes and look for the invisible mass between galaxies that may bind the great clusters together, then black holes can be of more help to us, though the same sort of constraints apply. For instance, if the nearby Virgo cluster were dominated by huge black holes of galactic mass, black galaxies if you like, they would have a destructive effect on the visible ones. We can see in Virgo only about ten galaxies which look tidally disrupted. In at least six cases we can attribute the disturbance to an obvious visible galaxy nearby. This suggests there are fewer invisible galaxies than visible ones in Virgo. However, a large number of lighter holes cannot be ruled out.

In summary, it is nigh on impossible to rule out with certainty black holes as principal agents in the missing light. They could exist in such a wide range of weights, between 10^2 and 10^6 solar masses, that any single argument can hardly expect to dismiss them all. No massive holes are so far known to exist, and no convincing reasons have been given as to why they should form in the sort of numbers and the sort of places required to explain the missing light. While black holes may be the natural end point of evolution for some visible bodies like massive stars or galactic nuclei, such evolved holes ought to be far too infrequent to explain the missing light. And when massive pre-galactic lumps collapse, both theory and observation suggest they collapse not into holes but fragment into smaller bodies like stars which can exist and shine for such a very long time that we would notice them today. But if we are entitled to skepticism about large numbers of black holes, we have to admit likewise to such an abysmal ignorance of the way in which other structures like galaxies have come into being that we can agree with Bernard Carr who has said '. . . there is as yet no reason for the quest for black holes to be turned into an inquest.'

So far as the missing light around galaxies is concerned we now appear to have run out of ideas. None of the currently fashionable theories seems to fit satisfactorily. And yet we cannot dismiss the observations. The stability and rotation curves of spiral galaxies call for two to three times the mass we can presently see. Measurements taken further out of satellites, of binary companions and of whole groups call for 90 per cent of the mass to be invisible. This is such an unsatisfactory situation that perhaps we should examine the proposed explanations more charitably. Perhaps we have pressed some of our objections too hard.

We know directly so little of the dwarf stellar population of other galaxies, or indeed of our own halo, that it is conceivable that the real, as opposed to the theoretical position, is not so bad as we have supposed. We have argued that a dwarf population will make itself visible, as it does in the immediate neighbourhood, by a small admixture of more luminous solar-type stars. This *could* be incorrect if the formation of such heavy stars is completely suppressed in the locations where Γ is high. We know too little of the star formation process to be very dogmatic about such suppression, even if it does seem very ad hoc. The real difficulty for the dwarf hypothesis is the stickiness it necessarily implies for galactic haloes. Someone has to show that the stickiness calculations are quite wrong before I would be prepared to resuscitate the dwarf hypothesis.

Apart from the present lack of any convincing evidence that neutrinos actually weigh anything, we found against them chiefly because their velocities are so rigidly defined. If one selects a velocity (and hence a mass) appropriate to keep them in haloes and so explain the missing light, they are then too slow to explain the mass discrepancy in large clusters, and too light to make up the missing mass in the Universe as a whole. But we may have been asking too much of them. Perhaps it could be that the three types of neutrinos (electrons, muon and tau neutrons) all have different masses, each appropriate to a different problem. Neutrinos then, even if they are beginning to smell a little, are not entirely dead.

The black hole hypothesis is so vague, and the range of possible masses and origins so wide, that a categorical refutation may never be possible. However, according to some eminent authorities an irrefutable hypothesis cannot be considered as 'scientific.' Be that as it may, black holes will, if anything, give stickier haloes than dwarfs.

If the Universe is closed then whatever the explanation for the missing light, we are faced with the even greater mystery of the missing mass. The missing light problem calls for galaxies with mass-to-light ratios Γ in the range of 30 to 100, implying a cosmic density parameter Ω of between .03 and 0.1. Pressing Ω to the closure value of 1 implies an even vaster reservoir of invisible mass.

If the reservoir is composed of some stuff 'Y' the question which comes to mind is whether Y is just more of the very same X we invoked to explain the missing light. It seems rather likely that X and Y are distributed very differently in space. Galaxies are heavily clustered in space so X must be clustered with them. But if the much larger mass of Y were clustered in the same way, the cluster masses and gravitational fields would be so large that galaxies would be pulled about at random speeds of about 800 km/sec. We only find such speeds in great clusters which contain less than 10 per cent of all visible galaxies. The majority of isolated galaxies and galaxies in small groups have much lower random velocities around 200 km/sec. So if Y exists it must permeate space as a whole, residing predominantly in the great voids between clusters and groups. And whatever Y is, it must be very dark indeed. The large mass combined with the upper limits measured for the extragalactic background light calls for a Y with a mass-to-light ratio Γ of 1000 or more. Among the various candidates for Y are very hot intergalactic gas, neutrinos, black holes, gravitational waves and intergalactic stars. But to all these candidates more or less cogent objections can be raised.

A handful of truly intergalactic stars are known, mostly discovered by accident. For instance, in a recent study of hundreds of thousands of faint red stars at Edinburgh, most of which are nearby red dwarfs, one star was found with colours suggestive of a luminous giant. An optical spectrum taken later on confirmed its giant classification. To appear as faint as it does, the giant must be way outside our Galaxy, or indeed any galaxy that can be detected. It appears to be a truly isolated intergalactic star.

Some intergalactic stars are expected. As galaxies brush by one another some few stars should be thrown outward by the interacting gravitational fields. And indeed, bridges or filaments of stars can sometimes be seen stretching from some spirals to their neighbours in space, while splash-like rings of stars have been discovered round some ellipticals which suggest they have recently undergone a violent merger. In the past when galaxies were much closer together such encounters and collisions would have been more frequent leading to a steady build-up of the intergalactic stellar population.

Such ejected stars, however, cannot be plausible candidates for Y. Their mass-to-light ratio Γ ought to be 10 or less like the population from which they are drawn. This is a long way short of the Γ equals 1000 needed for Y. And all the objections to dwarf stars as candidates for X apply with even greater force to Y.

But what of a population of native intergalactic stars that were born and have always existed out there? Here we must appeal to theory. The big bang model tells us that when the pre-galactic medium was dense enough for star formation to take place, the radiation pressures were too intense to allow of it. And by the time the radiation had cooled sufficiently the intergalactic gas would have been too diffuse. If stars formed out there at all they ought to have formed in clusters of a million or more. Since no such clusters are found it is generally concluded that Y cannot be made up of intergalactic stars, or indeed of any dense bodies including black holes, that are less than a million suns in weight. All this presupposes that our rudimentary understanding of how stars form is more or less right.

Equation 11.1 tells us that if Y is predominantly made up of massive neutrinos then their present-day speeds would be no greater than 10 kilometres a second. Such low-speed entities would naturally fall prey to gravitational fields and be snatched into galaxies and groups, just where we do not want them. So massive neutrinos are probably ruled out. Equally, the coherent objections to gravity waves have been rehearsed in Chapter 10.

This leaves only a very hot, very diffuse intergalactic gas as a plausible candidate for Y. Such gas, you will recall, will not radiate much because the atoms would be too far apart for many atomic collisions to take place. And what radiation there is should appear not in the visible but as a diffuse isotropic source of X-rays. The idea of such a lot of uncondensed gas is appealing because our knowledge of gravitational condensation tells us that it is generally very inefficient. We should not be surprised to find that when the galaxies condensed out, 99 per cent of the gas was left in its uncondensed pristine form. Such gas would detectably absorb light passing through it unless it was more than a million degrees hot.

It is encouraging therefore to find that the isotropic background of X-radiation discovered by Giacconi and his colleagues has a spectrum well fitted by a gas at a temperature of 400 million degrees centigrade and with a density sufficient to close the Universe. Such a hot all-pervading medium ought to interact with the cosmic microwave background distorting its spectrum towards long wavelengths. Controversial and very difficult measurements made from a balloon hint that just such a distortion may be present. But more reliable satellite observations will be required to settle the issue.

We have already agreed that, due to the universal expansion, undisturbed intergalactic gas remaining from the big bang ought by now to have cooled to only one hundredth of a degree above absolute zero. If we require it to be at 400 million degrees today then we have to think of an enormous source of energy which could have reheated this gas. Quasars are a possibility. They were commoner in the past than they are now. And they belch out copious amounts of ultraviolet radiation which would be absorbed and heat up any surrounding gas. The snag is that the quasars we can count fall short of providing the necessary heat by a factor of a thousand or more. If all galaxies once underwent quasar-like outbursts, as is conceivable, then that might have done the trick. If nascent galaxies gave birth to a first generation of massive stars, these stars would have generated the necessary ultraviolet output before collapsing into presently dark tombs. Perhaps their tombs might account for the missing light in haloes.

It is all just possible. But consider the snags. The hot X-ray background quoted as evidence of hot intergalactic gas can be explained quite naturally in other ways. For instance, quasars and active galaxies are themselves such good X-radiators that they can account for most, if not all, the X-ray background on their own. Also, as we shall see in the next chapters, some cool clouds of intergalactic gas have now been found. They would not have survived if surrounded by a very much hotter high pressure medium.

Although enthusiasm for hot intergalactic gas is still sustainable, it calls for a fair bit of whistling in the dark. To quote from a recent review on the subject by an expert, Joe Silk: 'the search for traces of the Intergalactic Medium has been long and often fruitless. Reputations have been made or broken, often only for what has finally proved to be a new, perhaps stronger, upper limit on what might or might not be present in intergalactic space. Radio, optical, space-ultraviolet, X-ray and gamma-ray astronomers have all confronted the challenge, and all have failed to find the elusive, uniform, intergalactic medium.'

At the end of all this argument and counterargument we find ourselves at a loss. Whilst we have convinced ourselves that the missing light definitely exists, we cannot explain it. All the ingenious theories put forward to account for it seem to suffer from grave defects of one sort or another. Such survivors as exist are either implausible or unverifiable by present-day means. At the same time there is a natural reluctance to admit that we have no idea as to the identity of 90 per cent or more of the substance of the Universe. With nothing to put in their place, astronomers are reluctant to discard such theories as they have, unhealthy though they may be. The subject is a large one and no one would claim to be an expert across the whole field. While the neutrino expert may have convinced himself that neutrinos just will not do, he is not so certain about dwarfs. And vice versa. We are like a team of specialists standing around a corpse that is beginning to smell. No one is quite prepared to sign the death certificate just in case there may still be signs of life in some part of the anatomy outside his range of specialized knowledge.

The missing mass problem is not so bad. Until observations yield a more certain value for Ω we are free to assume that the missing mass does not exist. We may be left uneasy that Ω appears to be close to 1 without quite making it, but it is an aesthetic unease, not a scientific argument.

My view, for what it is worth, is this. By all means let us continue to test and probe the current theories. But let us not pretend that the only or most likely explanations will certainly be found amongst their swaying ranks. Let us look further afield, as astronomers have had to do so often in the past. We may have to explore some unlikely byways and possibly upset a cherished applecart or two. Even if we do not find the holy grail we are certain of some intellectual fun along the way.

Illusions and Assumptions

$\mathbf{W}$E have at great length discussed the COWDUNG ('Conventional Wisdom of the Dominant Group') which supposes there to be a vast preponderance of undetectable mass in the Universe. One may, like many astronomers, find this convincing, or one may be more skeptical and doubt the evidence of the emperor's new clothes. What one cannot do is ignore the evidence: spiral rotation curves *are* flat, and cluster velocity dispersions *are* far too high. Those who cannot stomach the conventional hypotheses must look for alternative and heterodox ways of explaining the observations. In this chapter I want to discuss two heretical ideas on which I am working and which seem promising to me.

In speaking of the luminosities of galaxies and of their contribution to the total visible content of the Universe, we have glibly skated over some controversial questions. How do we define the total luminosity of a galaxy? How, in an ideal Universe, could we measure it? And to what extent do our presently imperfect techniques allow us to approach a reasonably correct answer? Is it conceivable that we have underestimated the true luminosities of galaxies by as much as a factor of 10? If so, that would go a long way to remove the problems both of the missing light and the missing mass.

That these questions are not entirely rhetorical can best be seen by exchanging our present location for an equally likely position in space. Suppose the solar system inhabited, not the outskirts of a spiral like the Milky Way, but the inner regions near the centre of an elliptical galaxy. Since many of the sun-like stars in the Universe reside in just such ellipticals this is not such a fanciful exchange to make. Seen from our new location the night sky would be ablaze with giant stars ranging in colour from furnace orange to pigeon's-blood red. There would be in our sky no less than 2000 stars brighter than Sirius, with the brightest being 300 times brighter than any star we can presently see. In all, there would be 3 million such giants visible to the naked eye, 30 or so in each patch of sky no larger than the moon's disc. You can add to that about 20,000 dwarfs and a background glow from a myriad other stars too faint to be seen as individuals. And just as it does now with moonlight, the Earth's atmosphere would scatter this brilliant harvest of light into a diffuse glare stretching from horizon to horizon.

Stellar astronomers would of course be ecstatic but extragalactic astronomy would scarcely exist. The ubiquitous stellar glare, two or three times more intense than the brightest full-moon-lit night, would swamp the faint signals from external galaxies almost completely. The existence of spiral galaxies would be unsuspected. The faint outer margins of external ellipticals would likewise be veiled in the atmospheric light. The only extragalactic objects still marginally visible from the ground would be the intense cores of other ellipticals like our own. Outside our immediate environs the Universe would appear to us as a barren and almost empty void.

How fortunate, you might think, that we are not so situated. But this must be a relative matter. While we are not so blinded as our contemporaries in an elliptical, we are nonetheless imprisoned in one of the spiral arms of a giant nebula. Who can say how much more there would be to see were we truly liberated from the bonds of neighbouring light and floating darkly in the intergalactic void. What faint and filamentary structures we might map; what wonders previously unseen might delight the eye and complete the tapestry of the Universe. It is as if we are imprisoned in a lighted cell. We can stand on tiptoe and peer out to the darkened world outside. We can see the street lamps and the pavements beneath. We can see the headlamps of passing cars and here and there a lighted window. But the remaining and greater part of the world can only be guessed at. Those leaves waving in the lamplight suggest the presence of a darkened tree beyond. That window high above betokens a substantial building: but how big and of what shape?

What I want to suggest is this. There may be a good deal more luminous structure to the Universe than we can presently see, limited as we are by the glare of our immediate surroundings, and in particular of our atmosphere. There may be an abundance of faint galaxies and other structures that we cannot presently detect. And even those we can see may extend far out beyond the limits to which we can currently measure. This, of course, is an easy claim to make, and not particularly original. What we must do is look for evidence. Since, in the nature of things, the evidence cannot be direct, we must look for the circumstantial variety. Not only does it exist, but I claim it is rather convincing.

Let us begin by looking at what is known of the luminosities of elliptical galaxies. We shall come to spirals later on. Most of what we know about them comes to us from the use of photographic plates. Such plates have a very limited storage capacity. Even exposures taken on a telescope at a dark site and on a moonless night cannot be continued for more than an hour or so. After that, the plate saturates or blackens all over, largely because of zodiacal, atmospheric and galactic light. Any source too faint to impress its image on the plate during that time will remain undetected. When one considers that a plate records only about 2 per cent of the light which falls upon it, the other 98 per cent going to waste, one realizes how severely handicapped the astronomer is in his work.

The centres of ellipticals, being 250 times brighter than the sky, quickly show up on a photographic plate. Indeed, for exposures continued beyond a few minutes the central areas will be burned out. The outer parts, on the contrary, fade away and merge so imperceptibly with the night sky that it is *impossible* to say where the galaxy ends (see Plate 8). And all this despite ingenious techniques which enable us to follow the profile outward to the level of less than half a per cent of the night sky. You might suppose that at such low levels it does not really matter where the galaxy terminates because the extra light involved cannot add significantly to the total. Not so. The falling intensities are compensated by the larger areas from which they come. Since the outer parts can neither be detected nor ignored astronomers are forced to 'extrapolate,' or in plain English, to guess. They measure the falling intensity out as far as they can and then assume that it continues on in the 'most natural' way. What they do in practice is to fit smooth mathematical curves to the reliable inner observations, hoping, with no good cause, that these curves remain a close fit much further out. The fiendishly difficult part is to estimate the level of the foreground night-sky overlying the galaxy, for this must be subtracted from the observations. Set this level half a per cent too high and you artificially amputate the galactic exteriors; set it half a per cent too low and the galaxy will appear to be of infinite extent and luminosity. Because the foreground light is often uneven, and because neighbouring galaxies interfere with one another, galaxy photometry is necessarily something of a black art.

Edwin Hubble made the first systematic study of elliptical profiles in the 1930s. He found them to be generally very smooth and regular, fitting well to the simplest form of mathematical curve in which the measured light intensity falls off as the square of the radius out. Although the empirical fit between curve and observations appears to be excellent, the implications are ridiculous. If you add up all the light under Hubble's curve as it continues outward, the total luminosity does not converge to a finite limit. Extrapolated too literally, Hubble's work implied that every elliptical galaxy was infinitely luminous with most of the light coming from very far out where the local intensity has fallen well below the level of the night sky. Clearly then at a low enough intensity or a large enough radius, Hubble's law must break down. Precisely how and where it breaks down will determine the true total luminosity.

The modern attack on the subject has been led by Gerard de Vaucouleurs. Using slightly better plates and more elaborate techniques for taking care of the sky, de Vaucouleurs finds profiles which are better fitted by a logarithmic curve which *is* convergent and yields a finite luminosity.

If the de Vaucouleurs curve truly fits all the way out then the total luminosity of an elliptical can be estimated from the values of two measurable parameters: the central intensity or surface brightness $I(o)$ and the radius r_{20} at which the intensity has fallen to one twentieth of its central value. Since the central intensities are generally 100 times higher than the sky background both these parameters are readily accessible to measurement. The total luminosity L can then be computed as $L = I(o)r_{20}^2$. So far as we can see the de Vaucouleurs curve fits the profiles better than any other, but the whole procedure relies on two radical assumptions. It assumes firstly that all elliptical profiles follow the same canonical form; secondly that the canonical form continues smoothly outward to light levels well below the present feasibility of measurement. It is as if one were to say that all icebergs have the same form and that therefore the total volume of any berg can be estimated by measuring its peak height above sea level and the width of the peak twenty feet below the summit.

Insofar as we can check them, the de Vaucouleurs assumptions seem to work rather well. However, we would be much more confident if we really knew how elliptical galaxies 'worked.' But the fact is we cannot even agree as to their true 3-dimensional shapes. Some astronomers believe they are oblate or top-shaped, and supported by rotation. Others think of them as rugby balls (prolate) which are supported by anisotropic pressures left over from the process of formation. As we see them projected on the sky, both shapes would look much the same, so it is not presently possible to distinguish for sure between the two alternatives. It is a measure of our ignorance and a warning that dogmatism as applied to any aspect of ellipticals is surely out of place.

Whatever our sense of unease the de Vaucouleurs technique is at least a self-consistent, practical procedure both for defining and measuring the luminosities of elliptical galaxies and the luminosities we shall henceforth quote have been measured by means of it.

In 1957 Fish set himself the arduous task of measuring the de Vaucouleurs parameters for a sample of about fifty ellipticals which were deliberately chosen to span a wide spectrum of intrinsic properties. They ranged from supergiants in the heart of the Virgo cluster to the dwarf companions of Andromeda which are no less than 10,000 times fainter. That is to say they are *intrinsically* fainter but not necessarily apparently so, because they happen to be so much closer. Why he was interested in the sample need not concern us here; we shall be content to know that he measured the central surface brightness $I(o)$ for each galaxy and the de Vaucouleurs radius r_{20} (for which purpose he also had to know their distances). He then calculated an absolute luminosity L in the usual way from $L = I(o)r_{20}^2$. The result was an unusually homogeneous data set on which he based an ingenious and elaborate theory as to the way ellipticals might have formed.

In 1975 I found myself puzzling over Fish's very interesting paper. But the more I looked at his theory the less I liked it. Nevertheless the data were clearly trying to tell us something interesting. But what? When in doubt it is often a good idea to look back to the basic observations. Unfortunately these did not appear in Fish's paper. However, the observed values for $I(o)$ and r_{20} could be reconstructed from the figures he actually gave. So I proceeded to work them out, and as I did so something remarkable began to emerge. To within rather close limits all the galaxies in the sample appeared to have the identical central surface brightness $I(o)$.

I found this remarkable because in so many other ways the galaxies differed radically from one another. Why should a dwarf containing only 10^8 stars and a giant with 10^{12} have exactly the same value for I(o)? To return to our iceberg analogy, it was as if we had found that all the floes in the sea, from the smallest to the largest, had identically the same height at their summits.

Coincidences like these generally arise from one or two fundamentally different causes. There may be a physical reason of fundamental significance at work (as Fish had surmised) or else the insidious effect of observational selection may be responsible. That is to say, in picking his targets an observer may unwittingly select a class of objects with atypical or peculiar properties. This is all too easy a mistake to make and one that has plagued astronomy down the centuries. For instance, we saw in Chapter 2 that the sample of apparently brightest stars consists largely of very rare supergiants.

I wondered, therefore, about Fish's sample. Had he accidentally selected out only galaxies with a particular central surface brightness I(o)? Looking through his observing list, I recognized most of the galaxies as well-known brutes that are both apparently bright in the sky and of large apparent angular size. This made sense, for to apply de Vaucouleurs technique with precision Fish needed the largest and brightest galaxies he could find. But it gave me cause to wonder. Could it be that in picking out galaxies of large apparent radius he had inadvertently selected out objects with a preferred value for I(o)? I began to do some calculations.

The apparent radius r(ap) of a galaxy is a parameter quite separate from the de Vaucouleurs radius r_{20}. Whereas r_{20} is measured at an intensity level of 1/20th of the peak intensity I(o), r(ap) is the full distance out from the centre to where the galactic intensity appears equal to that of the surrounding sky. On a photograph a galaxy will *look* as big as r(ap) not r_{20}.

What I suspected was this. In sensibly picking out galaxies of large apparent size Fish may have unwittingly selected a sample with just that central intensity I(o) calculated to give them the largest apparent size on photographic plates.

Plate 8 lends plausibility to this idea. It shows two plates of a pair of companion ellipticals with slightly different peak intensities or I(o)s. On the upper plate the left hand elliptical is clearly the larger and apparently the brighter of the two. In the lower plate, which is a longer exposure but otherwise identical, the situation is reversed. The right hand galaxy, having a lower I(o) but a larger r_{20}, reveals itself only when more light has been stored upon the plate. Thus the different sizes of galaxies on photographs depend upon the I(o)s and the finite storage capacity of the photographic process.

Since we abstain from mathematics here, we shall look at the situation pictorially. Fig. 18 represents three galaxies as icebergs with the upward dimension corresponding to light intensity and the sea surface dimensions corresponding to the plane of the sky. In such a representation the total luminosity of a galaxy corresponds to the *Volume* of the equivalent iceberg, and sea level corresponds to the intensity of the sky. The different pinnacle heights represent the central intensities $I(o)$ while the de Vaucouleurs radii r_{20} are in each case measured at a fixed height below $I(o)$. The observer will, of course, see the galaxies (icebergs) in plan view from above (see projection at bottom of the figure). The apparent radii $r(ap)$ he will see are the radii of the bergs at sea (sky) level.

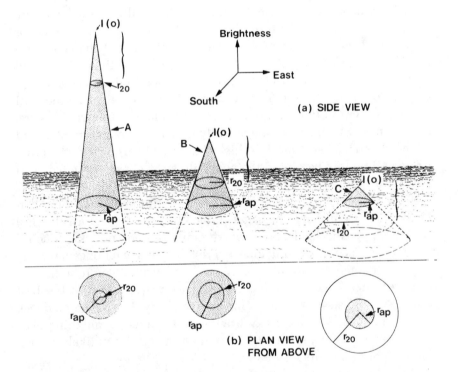

Fig. 18 Galaxies as icebergs. The three icebergs (galaxies) are of equal volume (luminosity). The dotted regions mark the volumes (luminosities) visible above the 'sea' (sky). The remaining invisible volumes (luminosities) are shown lurking beneath the surface (sky). Notice that the observer looking down from above will see B as apparently larger than either A or C, which have different heights (central brightness) $I(o)$. Thus in looking for apparently large icebergs (galaxies) the observer may be inadvertently selecting in favour of particular values of $I(o)$.

Galaxy A is an apparently small brilliant object with most of its light visible above the sky. Galaxy C appears as a small dim individual because most of its light is hidden below the sky; indeed if its I(o) were dropped much further C would disappear altogether. Galaxy B with an intermediate value of I(o) appears to be the largest of the three, and yet all three galaxies (bergs) are identical in total luminosity (volume). If the most spectacular looking galaxies in the sky are not the brightest but merely those with a felicitous value I(o), then what of all the so-called 'dwarfs?' Some might be giants crouching almost below the level of the sky, i.e. extreme examples of C in Fig. 18. Others, like A, might be compact, brilliant objects with their images largely burned out on photographs; extreme examples might even masquerade as stars. The human eye scanning the photographic plate would then have to be thought of as a most unreliable witness where elliptical galaxies are concerned. It would confuse size with luminosity, take giants for dwarfs, and select outrageously in favour of superficial characteristics like central surface brightness. It could never overestimate but could disastrously undervalue the true luminosity of ellipticals. The net up-shot would be that astronomers would always underestimate the total luminous contribution of any elliptical population. In great clusters, for instance, which are so dramatically underluminous ($\Gamma = 500$) and which are composed almost exclusively of ellipticals, part of the problem could simply be due to the above illusion, or 'iceberg effect' as I like to call it.

But I am rather racing ahead. All we have done so far is show that Fish's galaxies all have the same surface brightness with a value I(o) which happens to lend them a maximum size upon our sky. It could of course be a colossal coincidence. Providence may for once have just arranged matters in an improbably felicitous way. It could be that all giant ellipticals have so composed themselves as to make the best impression when looked at in our parochial sky. For the coincidence *does* depend on the local sky brightness as much as anything else. While such a Panglossian explanation seems wildly unlikely to me, one cannot rule it out absolutely.

On the other hand, if there were anything to this iceberg illusion, it seemed to me that it should apply with equal force to spirals as it does to ellipticals. Let us then briefly review the problem of spiral luminosities.

Spirals are slightly more complex in that they are composite objects with an inner elliptical core and a flat outer disc with superimposed spiral arms. Because, despite appearances, the disc provides about 90 per cent of all the light, we shall simplify considerably by ignoring the other components. If we look at such a disc more or less face-on, and we consider only face-on spirals for now, then it looks rather like an elliptical galaxy. The centre is bright while the outer parts merge imperceptibly into the background sky with no apparent outer boundary. Once again, most of the light comes from the faint outer regions so that in estimating the total luminosity we have to struggle with all the former problems of sky subtraction, convergence and extrapolation. The profiles are found to converge to a definite limit more quickly than ellipticals and to fit a different mathematical curve. According to de Vaucouleurs, however, each such curve can still be described by the two parameters of central intensity I(o) and radius r_{20}. Once again the total luminosity can be computed as $L = I(o)r_{20}^2$.

If the iceberg illusion works for discs, I expected to find that all the biggest and best known discs would have the same peak intensity I(o), calculated to give the most spectacular appearance in the sky. Because their light profiles follow a different mathematical curve from the ellipticals, I was able to show that the optimal I(o) for discs is 650 times fainter than it is for ellipticals. And these expectations are borne out exactly in practice. Indeed, in 1970 an Australian called Ken Freeman had collected together all the best measurements of about 50 nearby discs, and he had pointed, without explanation, to their anomalously uniform I(o)s. He was rightly puzzled because his galaxies in the sample were in every other way a very mixed bunch. Some were giants, some were dwarfs, some were richly cloaked in gas and dust, others naked. There were blue ones, red ones, galaxies with and without spiral arms, some were even classified as irregular. Moreover the individuals had 'been selected and measured by different astronomers for different purposes. Two things only they had in common: large angular size on the sky and a uniformity of peak-intensity, even more striking than I had found among ellipticals.

I was convinced then and I remain convinced now. The hand of coincidence can be plausibly stretched only so far. Here we have our 100 best studied galaxies nearly all exactly conforming to the prediction of iceberg theory. On grounds of probability alone I think we must take the iceberg illusion very seriously and consider its implications.

When we examine a group of galaxies on a plate, our eye is naturally drawn to a few spectacular giants which seem to dominate the rest. This, I am suggesting, is an illusion. And if we go on to assume that the light of the group comes largely from these few illusory giants, then we shall seriously underestimate the total luminosity. It follows that when we have measured the group mass by separate dynamical means we shall come to an estimate of the group mass-to-light ratio Γ which is too high. In other words the iceberg illusion may partially account for the anomalously high values of Γ found in many groups and in all great clusters.

But I do not wish to be carried away. To claim that the illusion exists, as I do, is one thing. But to go much further and blame it for most of the missing mass is another. No matter how far we may be deceived into underestimating the luminosity of individual galaxies, we cannot conjure up indefinite amounts of extra luminosity without exceeding limits already set upon the integrated extra-galactic-background light. The measurements allow for five or at most ten times more light than we presently attribute to galaxies. Taking the upper permissible limit of 10, and convolving it with an average mass-to-light ratio of 10 for all galaxies, only yields an Ω(vis) of between 0.1 and 0.2. It may be an important part of the story, but not the whole of it.

Even if the light radiation from a galaxy is swamped by our atmosphere, it does not follow that its radiation at other wavelengths will be similarly disguised. Hunting for submerged icebergs at other wavelengths will be an obvious test of the whole theory. So far the evidence is contradictory. At the 21 cm radio wavelength where cold hydrogen, which is a substantial component of many spirals, radiates, very few icebergs have turned up, though there has yet to be a systematic search. On the other hand, X-ray pictures of clusters reveal some pockets of X-radiation surrounding prominent visible galaxies, and other pockets at places where no galaxies are to be seen.

Other clues are more subtle. The recent discovery that the internal velocities of all galaxies are proportional to the quarter power of their luminosity can be explained by the iceberg theory, but not, so far as I am aware, by any other. Invisible icebergs lying in the line of sight to distant quasars should show up as absorption features in the quasar spectra. Far more such features are indeed detected than we can account for with visible galaxies, and this is promising.

My colleagues and I are hunting for icebergs right now. Ed Kibblewhite at Cambridge has recently completed a marvellous machine which scans whole photographic plates with a laser. Being computer controlled, it allows us to measure hundreds of galaxies at once without human intervention, selection or bias. With its help and with the help of a new detector that is about to replace the photographic plate (see next chapter), some truly substantial icebergs will, we hope, be raised from the deeps of the sky. Unless we can do this, most astronomers will remain as skeptical about what they cannot see as they are today.

The current situation seems to be this. No one disputes that the iceberg illusion can deceive us over both elliptical and spiral galaxies. But whether it is responsible for hiding significant amounts of light remains as a fascinating problem which we are trying to solve. If galaxies are born with a wide range of I(o)s then we are certainly being deluded into an underestimate of their total light. But if the I(o)s are intrinsically all rather similar, as current interpretations assume, then the illusion cannot be responsible for much hidden light and mass. The real uniformity of the I(o)s then cries out for an alternative physical explanation. And the fortuitous coincidence of these I(o)s with the values which would show galaxies of both types to their best advantage in our earthly sky would have to remain a coincidence of such felicity that even Dr. Pangloss might have blushed.

Missing mass and missing light problems both arise from discrepancies between the luminosities of structures as observed, and their masses inferred by dynamical means. We have found good cause to doubt our luminosities; how certain can we be that the masses are right?

In weighing an astronomical system we generally begin with observation of its internal velocities; velocities which would quickly lead to structural dissolution if unchecked. To stabilize structure we invoke the existence of attractive gravitational fields, which we assume to rise in direct proportion to the mass involved. From the radius of the structure and the observed velocities we then calculate the necessary stabilizing mass by means of the universal law of gravitation. Where the mass so arrived at is 'unreasonably' high, as it so often is, we attribute any discrepancy to the existence of invisible or hidden material of various kinds.

Thus the weight of a spiral is inferred from the amount of gravitation required to hold its spinning parts together. Since the visible mass is apparently not sufficient we are forced to the notion of a dark halo. In ellipticals and clusters the fissiparous motions are not rotational but random, although the results are effectively the same. To supply the necessary gravitational forces we must often invoke otherwise unreasonably large amounts of massive invisible X.

Between observation of the velocities and computation of a mass, two conceptually distinct steps are involved. In the first, we take the laws of motion for granted and calculate the gravitational binding force necessary to hold the structure intact. In the second we assume that the law of gravitation is known and work backwards from the necessary binding force to the mass required to provide it.

There are all sorts of reasons for believing that we understand the laws of motion. They are regularly tested in the laboratory to great precision and so there seems no point in questioning their universal validity. The law of gravitation on the other hand rests on far less secure foundations.

You will recall (Chapter 5) that Newton and Einstein have proposed two conceptionally quite different modes of describing gravitation: Newton in terms of force and Einstein in terms of spatial curvature. But for the practical prediction of celestial motion the two laws concur save in fields so strong that the escape velocities are of the order of the speed of light. The internal and relative velocities of galaxies are only one thousandth of that speed, so in galactic astronomy we shall stick by the much simpler Newtonian law to describe the effects of gravitation.

Newton's law states that any two bodies attract one another with a force proportional to the product of their masses, and inversely proportional to the square of their distance apart. As applied to the solar system it gives accurate results; but when applied to the scale of galaxies it always gives rise to anomalies of the same sort. The visible masses give calculated forces that are always too small to hold the structures together. And the larger the structures the more inadequate the Newtonian forces turn out to be. Within galaxies the forces are twice too small; within great clusters they are but one hundredth of the necessary magnitude. The conventional response is to postulate increasing amounts of invisible mass; the only alternative is to believe that on the galactic scale Newton's inverse square law must simply be wrong. And it is this heretical viewpoint that I want to discuss next.

We can begin by asking what sort of modification of Newton's force law might bring the observations to square with the visible facts. What we want are forces that are always greater than Newton's at galactic distances. In mathematical form the Newtonian force F between masses m_1 and m_2 a distance d apart can be written as

$$F = Gm_1m_2\left(\frac{1}{d^2}\right) \tag{12.1}$$

Now we could modify this by including an extra term so that the force becomes

$$F = Gm_1m_2\left(\frac{1}{dD_0} + \frac{1}{d^2}\right) \qquad (12.2)$$

where D_0 is some very large fixed distance. The extra term denotes a component of gravitation that does not die away as rapidly as the inverse square law at great distances. For instance d much less than D_0 the new term is relatively insignificant, and so can be ignored, and the force law reduces to Newton's well-worn form. But when d is as large as or larger than D_0 the new term becomes dominant and increases the force of gravitation between two bodies at long range, and this is just what we want. For instance if D_0 were much greater than the scale of the solar system there would be no call to modify Newton's very ac-curate planetary predictions. But if D_0 were of galactic size the new term would provide all the extra force necessary to hold a spinning galaxy together without calling for any invisible mass.

Now I am not suggesting for one moment that (12.2) is in fact the correct form, though it does explain the flat rotation curves of spiral galaxies rather neatly. I simply wish to demonstrate that there are alternative mathematical forms of the law of gravitation which can be set up in competition with the original proposed by Newton. It must be the object of observation and calculation to determine which form is most likely to be right. When Kepler and Newton first set out to explain the planetary motion, they set up and discarded many hy-potheses before they found the one which fitted the observations best. To apply blindly that law over scales for which it was never intended is bound to be a risky venture. The extrapolation involved is breath-taking, for a cluster of galaxies spans a distance no less than 100,000,000,000 times the radius of the whole solar system. Extrapo-lations of this magnitude are seldom reliable, for instance they do not work at all in atomic physics. Why then have astronomers stuck so pig-headedly to their inverse square law even when its predictions are so often and so widely at variance with the observations? The fact is that whenever the Newtonian law is applied over distances of galactic scale, it gives rise to such strange results that we need to call for large amounts of otherwise undetectable material to put things right.

In this instance pig-headedness has its reasons. While Newton's law may have its problems there are formidable obstacles to the discovery of an acceptable alternative. The astronomer is a passive observer, not an active experimentalist. Whereas a physicist can test his most exalted theories on the workbench no laboratory is large enough and no astron-omer lives for long enough to try out many quite plausible theories of gravitation. Suppose, for instance, we wanted to compare the two theories represented by (12.1) and (12.2). As we all know, Newton's theory (12.1) predicts elliptical orbits for the planets. Equation (12.2) predicts rosette-like orbits in which each elliptical circuit fails by a certain angle to catch up with its predecessor. The size of that angle depends on how big D_0 is compared to the scale of the orbit. A large D_0 yields a small angle, and if D_0 were of galactic size the angles would be far too small to be measurable. To all intents both laws would then yield indistinguishable elliptical orbits on the planetary scale.

For better tests we are forced to look at orbits of much larger dimension: at binary stars or galaxies, for instance. Unfortunately the periods of gravitational orbits increase steeply with their radii. To test for a D_0 of galactic size we would need to follow the progress of a pair of binary galaxies for a billion years or more. Our tens of years of observation leave us almost powerless to discriminate between one gravitational law and the next. When such impotence prevails the natural instinct is to cling to the familiar Newtonian law with a desperation approaching religious conviction.

Part of our reluctance to abandon the old theory springs from a more practical and less emotional cause. In his original *Principia* Newton was able to show a unique and remarkable attribute of his inverse square law: a hollow spherical shell of uniform material exerts no net gravitational force on an object floating anywhere in its interior; the pull in any one direction is exactly balanced by an equal pull towards the antipodes if the inverse square law is true. Consider now an isolated astronomical system, a cluster of galaxies for instance. The rest of the Universe can be thought of as a series of roughly uniform spherical shells surrounding it like the skin of an onion. Since Newton had proved that no one outer shell can affect the cluster it follows that the net gravitational effect of the rest of the Universe upon it will be zero. In other words the internal behaviour of the cluster can be considered in isolation from the Universe surrounding it.

This is an immensely useful simplification for it means we can get on and study individual sub-units of the Universe without first having to map and understand the cosmos as a whole. No other law behaves like Newton's in this respect. For instance, if the law represented by (12.2) is correct we have a forbidding situation; one so complex that it is hard to know where to start. If (12.2) is true then it is easy to show that the distant reaches of the cosmos, regions of which we have very little direct knowledge, will affect a galaxy more profoundly than its near neighbours in the clusters. While, however, this may be a severe practical difficulty it does not necessarily prove that the law is wrong.

Because of the relatively narrow space-time horizon within which he must operate, the astronomer is in a very weak position when it comes to testing the laws which may rule the Universe on its largest scales.

The hard distinction between a scientific theory and an article of faith is that whereas one is refutable the other is not. While Newton's theory of gravitation at long range is refutable in principle, in practice it is not. It belongs in a twilight world half way between religion and science. Its companions there include Darwin's theory of natural selection and Wegener's ideas on continental drift. They are barred from complete respectability not by inherent weaknesses but by the short lifespan of the human race. No one denies the importance of such invulnerable theories, as I shall call them. They tie together vast bodies of disconnected knowledge, they illuminate the darkness and they provide flexible new lines of thought and action. But so do religions.

There is a very real danger of becoming hypnotized by Newton's very beautiful construction, or by Einstein's variations upon it. No science can afford to confuse irrefutability with infallibility. The lines between reverence and credulity, between facility and fashion and between a hypothesis and a dogma can be very indistinct at times. If eternal vigilance is the price of freedom then unswerving skepticism is the price of scientific truth. I think we have to accept that our knowledge of gravitation at long range must necessarily be tentative and that open-mindedness all round should be the order of the day.

In the face of doubts on this scale there are several ways to proceed. One is ruthlessly to hunt down the prevailing dogma hoping to herd it into confrontation with the observations. And in searching for the missing mass and the missing light, that is what we are hoping to do. The second is to perfect our methods of observation and measurement to a degree that may catch out Newton's law. Our progress in this direction is spectacular: for instance, laser beams bounced off the moon can now measure the lunar position and orbit to within a centimetre or two. Third, we should actively look for new theories of gravitation. This requires not only energy and imagination but courage as well. The fertile young scientist still looking for a secure job may rightly shy away from an endeavour that could swallow up all his time and earn nothing more than the reputation for being a crackpot. And in this respect the sociology of science is badly letting us down. Science has tended to become as conservative as any other long-established profession. The well-beaten path to a university lectureship or some other permanent job is via a dozen or two 'sound' papers and not via a reputation for lunacy. And yet, in a sense, lunacy is what it is all about.

But theories can sometimes be overthrown from the most unexpected directions. There is great ferment at present in the field of particle physics. Grand unified theories are now the rage, and these have implications which reach out far beyond their original frame of reference within the atomic nucleus. It may yet transpire that gravitation can be explained in terms of something more basic. The precedent for this was James Clerk Maxwell's astonishing discovery that light is a manifestation of electricity.

The two ideas in this chapter are unconventional. But if they are not widely believed neither can they be dismissed as inconsistent with the facts. The observations of rotation curves in spirals, of binary galaxy orbits and of cluster velocity dispersions call for radical explanations of one kind or another. They no longer leave a safe position to which the conservative skeptic can retreat. If he does not like icebergs, will he be any happier with legions of unseen black holes? And if Newton's law is sacrosanct what alternative is he willing to offer?

Astronomy is now at a very exciting juncture. The observations we already have are so uncomfortable that we shall be forced to head off in one radical direction or another. Which way we go will probably depend most on the exciting new tools and techniques that will be coming into operation in the near future.

The Future

IN the past progress in astronomy has been limited very largely by technical considerations. The situation today has dramatically changed. Now that we have broken the bonds of gravity and atmosphere the technical possibilities are almost limitless and the pace of progress will be constrained almost entirely by financial and political decisions. At one extreme one can imagine whole colonies of astronomers living in space and equipped with the most marvellous tools for studying the Universe. Equally one can anticipate a climate in which research programs and universities are even further starved in the name of fiscal policy. In this chapter I want to take a look into what the future may hold for our problem. Deliberately I shall be conservative and confine attention largely to the next decade or so. In fiscal terms I shall be moderately optimistic: I shall assume that the developed countries will *continue* to spend on astronomy and space research about one tenth of what they currently spend on toilet paper.

Undoubtedly what we would most like to have for our problem is an accurate and unambiguous value for the curvature of space-time, or of its equivalent density parameter Ω. We want not an $\Omega(vis)$ limited to visible material nor an $\Omega(dynamic)$ limited to material clustered like galaxies. We want the global Ω which measures the full curvature of the cosmos taking account of all the contained mass, observable or invisible, clumped or unclumped. And we want a value free of uncertain assumptions about either cosmic evolution or Einstein's cosmological constant.

If we do not succeed in pinning down the global Ω, and no promising way of doing that is known at present, then the next most desirable number is Ω(dynamic) which requires the average density of all the clumped and clustered material in the Universe. Clumped materials give rise to gravitational fields which will accelerate galaxies hither and thither stirring them into random motions. The disposition of galaxies in space yields a measure of their clumpiness while their velocities over and above the smooth expansion flow will yield Ω(dynamic). Provided our neighbourhood is typically clumpy one does not need to work too far afield. All we require are accurate spectroscopic velocities and accurate distances for several thousand of the brightest galaxies. To find the velocities is simply a matter of dedicating enough observing time. But the distances are much trickier, and moreover they enter the problem in two ways. To get at the 3-dimensional clumping you obviously need to know the distances to galaxies supposedly inside or outside a clump. And to get a random velocity you must subtract from the measured velocity the expansion velocity which can only be calculated from the independently known distance.

Galactic distances are notoriously difficult to estimate, which lends importance to a new technique proposed by Brent Tully and Rick Fisher. These two young astronomers have suggested there is a one-to-one correspondence between the internal rotational speeds of spirals and their intrinsic luminosities. If this is true, and the indications look promising, then a relatively simple measurement of the rotation speed should yield the distance of a spiral with tolerable accuracy. They, and others, have undertaken a massive program to map the local gravitational field in this manner and there is every hope they will find at least an upper limit for the local Ω(dynamic) to which all astronomers can subscribe.

If either Ω or Ω(dynamic) can be measured with some certainty then mankind will have taken a historic step forward analogous to Eratosthenes' first computation of the size of the Earth. There have in the past been so many disappointments in this endeavour that I would not rate our chances of success in the coming decade at better than twenty per cent. If we could only find a clear cut way of fathoming galactic distances we might be able to see out of this particular wood at last.

The one really significant window of the spectrum through which we have never surveyed the Universe is the infrared, and cool bodies between temperatures of 10 degrees and 1000 degrees absolute (the Earth is at 300°) will radiate almost exclusively in this range. Since the radiation temperature of free space is only 3° most bodies which lack either a hot self-sustaining source of energy or a nearby illuminating source will cool and radiate, if they radiate at all, only in the infrared. Some people suspect that the greatest mass of the Universe will show up only when we can map this dim heat-radiation.

But the technical problems are prodigious. The atmosphere lets only a minute fraction of such radiation through, and even that which penetrates is admixed with a much greater confusing signal from the glow of our own atmospheric water-vapour. All bodies at earthly temperatures themselves radiate in the infrared. The Earth itself, together with telescopes, instruments, detectors and even astronomers, all spill unwanted radiation over the weaker signals from extraterrestrial space. Individual infrared photons are often too weak to be detected using conventional optical techniques, even when those can be stretched down into the infrared.

Nevertheless the potential prize is breathtaking. The infrared window contains ten octaves of wavelength between 1 micrometre and 1 millimetre as compared to only 1 octave for the visible. Infrared radiation can escape to us through clouds of gas and dust which entrap light and presumably hide many of the most exciting physical phenomena from our eyes. In a predominantly cold space it is natural to expect that material will largely radiate in the infrared. And finally in an expanding Universe the light from the very distant objects such as young galaxies will be redshifted into this region of the spectrum. It is, therefore, not surprising that astronomers are investing a great deal of effort into opening this particular window.

What we need is an observatory cooled down close to absolute zero and floating in space high above the atmosphere. To meet these objectives the Dutch and Americans, with some help from the British, have clubbed together to launch the first Infra-Red-Astronomical Satellite (IRAS) which was successfully launched in 1983. IRAS basically consists of a 70 cm diameter telescope and instrument package buried in a large flask of liquid helium designed to keep the satellite at the freezing temperature of 3 degrees absolute. Unfortunately the sunlight up there will boil off the helium, resulting in a design lifetime of less than 12 months. As it swings round the Earth the satellite will rake the Universe with dozens of heat detectors and telemeter the information back to Rutherford Laboratory at Didcot. Despite its modest size and short lifetime the experimenters expect to catalogue a million sources this way. The expected harvest includes red giants throughout the Milky Way, forty thousand external galaxies radiating through their interstellar smoke, and a considerable number of quasars and other exotic objects. The most exciting discoveries will, however, be those we cannot predict. Past history at other wavelengths virtually promises that IRAS will surprise us by trawling up new and unsuspected denizens from the deeps of the sky.

In the meantime we are doing our best to observe the infrared sky from the ground. Recently the UK has installed a 4-metre infrared telescope 15,000 feet up on the summit cone of Hawaii and the Americans have a similar 3-metre a few yards away on this choice site. Provided you know where to point them these telescopes can pick up faint infrared sources whose optical counterparts remain invisible. What they presently lack is an infrared analogue to the photographic plate, so they cannot easily be used to survey the sky in an unblurred way like the much more sensitive IRAS. However, it is rumoured that the defence industry has developed very expensive infrared imaging devices, and when these are declassified we can expect spectacular maps of the cool Universe even from the ground. Such ground-based telescopes work only because there are narrow windows for the transmission of infrared radiation through the atmosphere, particularly at short wavelengths; this limits their sensitivity. There is nevertheless a bonus. Compared to the optical the atmosphere seen in the infrared is less turbulent, so that finer details are less blurred by the so-called 'seeing.' Thus infrared telescopes ten or twenty meters across may one day picture the Universe with greater resolution than we presently enjoy in the optical. What we really need though is the money to put a more moderate, say 4-metre, infrared telescope into space where its sensitivity would increase a thousandfold.

As applied to our problem infrared astronomy may be able to detect galactic haloes made up of red dwarf stars. Also it will permit the spectroscopic study of massive stars during their red giant stage. If during this phase they blow off a great deal of matter, as is sometimes surmised, then they may decline eventually to become white dwarfs or neutron stars. But if insufficient mass is lost then one can expect black holes to be commonplace.

So many of our arguments have depended on our presently leaky understanding of how stars and galaxies form and of their distributions as between giants and dwarfs. That understanding will improve when we can use infrared radiation to see into the visibly obscured stellar nurseries in our own Galaxy, and outward to the highly redshifted regions where galaxies were presumably born. The light from nascent galaxies at a redshift of 10, say, will come to us shifted into the infrared.

The promise of infrared astronomy is almost boundless. But it will be, at least in the initial phases, an expensive science, and the prospects are clouded by various financial uncertainties. During the immediate future the missing mass enthusiast will be looking to see how much unsuspected cool material turns up on IRAS, and whether galaxies really have haloes made up of cool red dwarfs.

So much of our present understanding of the Universe comes to us from photography that it may seem curmudgeonly to list its faults and limitations. But astronomical photography is now a century old and it is about time we found something better. To begin with, photographic plates are inefficient and insensitive: even the best of them throw away 98 per cent of the incident blue light and virtually all of the red light. They are uneven, chemically dirty, and worst of all they have such a limited storage capacity per unit area that, despite their appalling insensitivity, they become saturated by the light from the very blackest part of the sky in less than an hour. Thus we are limited in the faintness and colour of the stars we can see, and we lose galactic haloes and other low surface brightness features against the sky. What we need is a more efficient, more red-sensitive device with unlimited storage capacity. Such is the CCD which is just coming into operation.

Among the many miraculous properties of the silicon chip is its unsurpassed sensitivity to visible and near-infrared light. Photons which are absorbed in the silicon structure give rise to precise electric charges which can be stored up or shuffled around electronically. Thus the pattern of light falling on the chip produces a sort of electrical photograph which accumulates in the silicon and which can be shunted out, amplified and displayed on a conventional TV screen. The sensitivity, which approaches 80 per cent, goes on far into the red where the photographic plate is blind. The storage capacity is effectively infinite because as it builds up, charge can be bled off and kept in a computer memory made of other silicon chips. Exposures can thus be continued indefinitely, though the net improvement over the photograph is less dramatic than you might think because of course the CCD collects more contaminating sky light as well as more starlight. Nevertheless perfected CCDs will be able to see features at least 30 times fainter than any we can presently detect. They are cheap and reliable and they are about to revolutionize astronomy. Their one present drawback is the chip-size which is limited to one or two centimetres, so that ordinary photography will still be needed for wide-field work.

CCDs will revolutionize the whole messy business of galaxy photometry. The existence of faint haloes, their continuance into the sky and their final convergence towards a measurable total luminosity are all problems for which the CCD is an ideal tool. Whether icebergs exist and whether light comes from within clusters but from without the perimeters of detectable galaxies should become answerable questions. In particular the dramatically enhanced red-response of these chips should allow us to see massive haloes if indeed they are made of faint red dwarfs.

The proportion of red and black dwarfs in the Milky Way will become more accessible to observation as will the population of extragalactic stars nearby. At last we shall pick up the light from very distant galaxies and quasars which has been redshifted beyond the visible range. This ought to provide a much firmer handle on galactic evolution, and therefore on the corrections that have to be made to some measured values of Ω. And CCDs mated to spectrographs will accelerate the measurement of the radial velocities required for mass-determinations.

As CCDs are being developed commercially for use in mass-produced, hand-held TV cameras they do not cost astronomy much. Another cheap product of commercial research which will benefit the observer enormously is the optical fibre being developed for telecommunications. Roger Angell working in Arizona has already used fibres to conduct the light from dozens of cluster galaxies into a single spectrograph in order to measure their velocities all at once instead of one at a time.

These, and other developments, particularly in the computer field, will certainly influence the design of future optical telescopes. The telescope is simply a device for collecting and focussing light into various instruments and detectors and the current design reflects the historical needs of the photographic observer. As photography is superseded, so the telescope can evolve to overcome some of its present limitations, notably its exorbitant cost. The massive blocks of glass together with necessary and expensive supporting structures can be divided up into a number of smaller, cheaper, separately steerable mirrors. The light from these may be combined using optical fibres or else each mirror can be fitted with its own CCD and the separately detected signals can be added electronically.

So many permutations are made possible by the computer, the fibre and the modern detector that astronomers can argue almost indefinitely about which is best. But the important conclusion is clear. Given the money, we could now assemble a giant optical telescope which by itself could collect ten times more light than all the telescopes so far built, working together. Such a giant could, for instance, measure spectroscopically the velocities of faint satellites and globular clusters lying far out in the dim margins of both spirals and ellipticals, determining once and for all the true extent and mass of galactic haloes.

We could expect such a giant telescope to impact all of astronomy because so many fundamental problems are insoluble simply for lack of sufficient light. In particular, studies of extended bodies such as galaxies, and spectroscopic researches on all classes of faint objects be they nebulae, quasars or stars, suffer acutely today from photon starvation. When one realizes that the last major up-grade in telescope size was taken in 1950 with the completion of the 5-metre on Mount Palomar (designed in 1928), astronomers can be forgiven a certain impatience in having to wait so long for something larger. The hiatus has already lasted now for thirty years and it will take us at least twenty more to plan, design and build a suitably ambitious successor.

Not all problems, however, will yield to sheer brute force. No matter how large a ground-based optical telescope may be, its resolution, that is to say its ability to distinguish fine detail, is entirely limited by turbulence in the atmosphere. Therefore the most exciting, most anticipated project presently on the stocks is the launch in 1986 of an optical telescope into space. Space Telescope, or ST as it is usually called, will be launched by the Americans on the Space Shuttle, with some help from the Europeans. It will be a free-flying remotely controlled telescope 2.4 metres in diameter. Equipped with CCDs, television cameras, spectrographs and a host of other equipment it will be virtually a complete space observatory which is expected to stay in orbit for at least fifteen years. Technologically it is a tour-de-force calling for state of the art developments in optics, guidance, detectors, electronics, control-engineering and mechanical reliability.

The promised performance is truly astonishing. It will image the Universe with an aerial detail at least one hundred times better than we can attain from the ground. In consequence star-like images will shrink to brilliant points of light with their contrast above the background sky so greatly enhanced that ST will detect stars a hundred times fainter than the faintest we can see from down there. The sky up there will in any case be always dark and cloudless and the instruments can feed on ultraviolet as well as visible light. We shall be able to see outward in space and backward in time ten times beyond our present horizon, and if we can disentangle the effects of galactic evolution and geometry the curvature of space-time should become apparent. ST may just be capable of detecting for the first time planets round nearby stars; it will, with ten times more precision, measure the motions of stars in the plane of the sky. I could go on almost indefinitely describing the opportunities we can expect because ST will see beyond our present capability as far as Galileo's first telescope saw beyond the naked eye. In 1986 there will be a breathless hush in the astronomical community as ST lifts off on its shuttle into space. Alas there is no back-up telescope if anything goes seriously wrong.

As far as missing-mass and missing-light problems are concerned we can anticipate help from ST in a number of directions. ST *may* just be able to measure the global Ω by observing distant supernovae either via the so-called Baade-Wesselink technique or by using them as standard candles. It appears that at maximum light the brightest supernovae all have the same intrinsic luminosity and as there is no reason to suspect evolution among supernovae we can apply the standard candle test to get at the spatial curvature. Nevertheless, the difficulties of finding the supernovae and of making the delicate measurements should not be underestimated.

We have mentioned already the possibility of assessing Ω(dynamic), which is a measure of locally clustered density, *provided* the distance to galaxies can be found with sufficient precision. ST is ideal for this purpose because it can see into nearby galaxies allowing us to recognize variable stars and other distance indicators. This, however, will be a slow business as hundreds of galaxies will need to be calibrated.

The high deuterium abundance in the interstellar gas has been used to argue for an upper limit on Ω of about 0.1 (see Chapter 7). But the deuterium measurements so far made pertain only to the local regions of space, in which case they may not be typical. The ultraviolet spectroscopy capability of ST should allow us to extend these measurements far further afield and if it turns out that deuterium has a universally high abundance this will confirm the low baryon density.

The standard cosmological tests, i.e. the standard candle and standard stick tests applied to distant galaxies, are, we know, beset with uncertainties about galactic evolution. Thanks to its wide wavelength sensitivity and its ability to map the morphological properties of galaxies out to high redshift, ST ought eventually to clear up our murky understanding of galactic evolution. And once we know how different types of galaxies change in luminosity and colour as they age, it should be possible to correct the standard cosmological tests unambiguously. Again this is a long-term problem requiring a good deal of hard work. Astronomers will probably begin by noticing how the properties of galaxies in great clusters change with time (i.e. with redshift). We expect the effects of galaxy mergers and of stellar energy decay to work in opposite directions, and we cannot presently tell which effect predominates.

With its capacity to pick up faint stars ST will help enormously to survey the true population of faint stars both in the Milky Way and further afield in galaxies of the local group. How many black, red and white dwarfs are there, of what does the old population on the halo and bulge mainly consist? Unfortunately there are no giant ellipticals in the local group, but for the first time we shall be able to make a census of the stellar population at least in the local dwarf ellipticals, which will tell us much of elliptical mass-to-light ratios, presently very uncertain. Above all we would like to know whether all the different populations exhibit the same distribution as between heavy and light stars. If there is indeed a universal mass distribution we shall be able to predict more readily what is going on in the distant giant galaxies which apparently provide most of the cosmic light and mass.

Whether ST will settle the contents of our own halo is open to question, though there can be no doubt it will set much tighter constraints. The small field of view of ST and the difficulty of distinguishing between foreground dwarfs and background and possibly intergalactic giants, will remain. However, ST will be able to map the position and velocity of very diffuse gas far out in the halo and disc of the Milky Way and of nearby giants like the Andromeda galaxy. It should be possible to see in the ultraviolet spectra of background galaxies and quasars, absorption lines due to this diffuse gas in the line of sight and hence to map Andromeda out far beyond the capabilities of the largest existing radio telescope. The motion of this gas will act as a sensitive probe of the gravitational fields and hence of the invisible mass which is supposed to populate the outskirts of giant galaxies.

However, whether any of these exciting new observations will eventually settle the main questions we have posed in this book seems doubtful to me. Even if we can finally measure Ω we shall still be left with uncertainties as to the distribution and nature of invisible material. At the same time ST is such a revolutionary instrument that we can anticipate a host of unexpected discoveries amongst which may be the one vital clue we need. We can be certain of only one thing: ST will cause some shocks.

If, as we currently assume, the hot big bang model of the Universe is correct then galaxies must have condensed out of an originally diffuse gas. Since both theory and observation agree that gravitational condensation is generally a most inefficient process, there are strong grounds for suspecting that most of the gas, perhaps the major mass fraction of the Universe, should still be drifting about in intergalactic space. Unfortunately, as we pointed out in Chapter 10, observations do not confirm this attractive idea, and the only unchecked loophole is the possibility that the gas is so hot, several hundred million degrees in fact, that it has remained invisible to contemporary observations. What future techniques could unveil this potentially significant intergalactic gas (IGG)? It ought to radiate in X-rays and it should interfere with microwaves from the cosmic background radiation, neither spectral range being accessible from the ground.

Although the cosmic background radiation was first discussed by radio astronomers, we now recognize that the vast majority of the radiation lies at far infrared and microwave frequencies absorbed by the atmosphere. Thus the most recent and most interesting attempts to observe it have been made either from balloons or from jet spy-planes flying high in the stratosphere.

The cosmic background radiation is fundamental to our understanding of cosmology. Not only is it the largest source of radiant energy, but it comes to us from so far away, from hot pre-galactic gas at a redshift of a thousand. Its existence is the best evidence for the big bang whilst its isotropy is the strongest argument for believing that the geometry of the Universe is basically simple and symmetric. Its inaccessibility from the Earth, however, has meant that so far we have been able to study it only in an incomplete and superficial way.

What we need is a comparatively modest (say 15 cm diameter) telescope in space fitted with instruments to detect radiation in the 0.1 to 1 millimetre wavelength range. Such an instrument, called COBE (for Cosmic Background Explorer), has long been designed in the States, but the money to launch it is still not forthcoming.

If there really is a lot of hot intergalactic gas about, the electrons in it will interact slightly with the cosmic radiation altering its spectrum. Already there are tantalizing suggestions from balloon-borne, and therefore not very reliable sources, that the spectrum is not precisely that of 3 degree cosmic radiation. It looks as if it may have been tampered with on its way to us from the big bang and the easiest way for that to have happened is interaction with hot intergalactic gas. But until COBE or some instrument like it is placed in orbit, we cannot be sure.

COBE should have a number of other interesting stories to tell besides. The movement of the Earth relative to the centre of gravity of the Universe will show up as a slight anisotropy in the cosmic radiation field: it will look hotter in our direction of travel. Crude ground-based measurements already suggest that the Earth, and with it the whole local group of galaxies, is falling roughly in the direction of the Virgo supercluster at about 500 kilometres a second. This high speed, which COBE could confirm to great accuracy, suggests that local gravitational forces, and hence masses, are possibly high enough for Ω to be near 1. Local gravitational waves will be detected as anisotropies of another sort, while gravitational waves way out in the Universe will show up as smaller corrugations of the surface brightness. Even the interpretation of the radiation as certainly coming from a hot big bang waits upon COBE. Only when we can see shadowing in the radiation as it passes through the gas in nearby clusters of galaxies will we be absolutely certain that the radiation is a cosmic phenomenon, and not something generated perhaps in the halo of our Galaxy. Hurry up, Congress, we need COBE rather badly.

Being probably so hot, the intergalactic gas itself will radiate mostly in X-rays, and although we have detected a strong isotropic X-ray background we cannot be sure as yet that it is not merely the sum total of signals from a lot of faint but discrete sources, possibly quasars. But X-ray astronomy is itself so strapped for cash that Riccardo Giacconi, its presiding genius, has just left the field for optical astronomy.

Giacconi's most ambitious effort was the X-ray satellite called Einstein which is now dead. Even with its tiny collecting area of 25 square centimetres Einstein sent pictures down to us of a whole new hot Universe. A larger X-ray observatory would be able to map out much fainter discrete sources, allowing us to judge if there is still room left for an X-radiating intergalactic gas. Also it would provide a spectrum with details of its temperature, density, clumpiness and composition. Indeed nothing could be more exciting than a big X-ray spectrometer in space. We have seen already the harvest of optical spectroscopy: the X-ray Universe, covering six octaves in frequency as against one in the optical, has no flying spectrometer of its own. British astronomers have offered to take a lead in getting one launched, but they have been turned down flat. It will probably be the middle 1990s before the Americans can afford a reasonable-sized X-ray observatory in space. What a bitter disappointment after the X-ray discoveries of the 1970s, especially as the X-ray telescope remains the best and perhaps the only hope of detecting black holes. Gas or stars falling into a hole should have X-radiation squeezed out of them. Apart from the controversial Cygnus X-1 type objects, we have so far had little success in this direction, but then our X-ray telescopes have been both short-lived and tiny.

Fortunately neutrinos can as easily be detected from the ground as they can in space, though 'easily' is hardly the appropriate word when discussing such wraithlike particles. Moreover the large and financially better endowed nuclear physics fraternity is just as interested in their properties as we astronomers are, so there is more hope of immediate progress on this front. Neutrinos, unlike black holes, can be manufactured in the laboratory, which should make the detection of any mass they might have so much easier. But 'easier' in this context is a purely relative term. To alter the words of Churchill, never in the history of scientific endeavour have so many laboured so hard to measure so little.

We really want the answer to two questions. How many neutrinos are there per unit volume of the Universe? And how heavy is each neutrino?

To the first question there seems very little prospect of a direct answer. Although, according to the big bang theory, about ten thousand billion neutrinos pass through each square centimetre per second, these big bang neutrinos will be of such low energy, and therefore such high penetrating power, that only a few will be stopped per day in all the oceans of the Earth. However, so long as we are convinced of the big bang theory then we can be confident that there ought to be 1 cosmic neutrino for every detectable photon, that is to say about 300 per cubic centimetre. For such a number to close the Universe each neutrino is required to have a tiny mass of about 10^{-31} gms, or 30 electron volts.

In nuclear laboratories all around the world physicists are doing their utmost to measure the neutrino's mass. They are already setting reliable limits of less than 50 eV, with one Russian claim for a measured mass between 14 and 46 eV. There are so many possible experiments, though none would appear easy, that if the mass is as high as 30 eV we can hope that someone will pin it down convincingly in the decade ahead.

Ultimately what we need is a telescope capable of surveying the

Universe for ultra high energy neutrinos. These will not be neutrinos from the big bang but from discrete sources like neutron stars and quasars. All you need is a cubic mile or two of water and thousands of photo-sensitive devices capable of picking up the brief flashes of Cerenkov light left by the particles formed when a high energy neutrino hits a water molecule. And such a heroic device is not out of the question. Some ingenious American physicists have suggested a project called DUMAND, short for Deep Underwater Muon And Neutrino Detector. In the clear waters of the Pacific, off Hawaii, where the sea is 5 kilometres deep, they propose to anchor a network of immense cables. At 50 metre intervals along these cables they would string porcupine-like bundles of optic fibres, each growing out of a photo-multiplier tube sensitive to light. As neutrino-induced particles flash by they trigger the sensors, which are all connected to a computer. The computer would trace back the track of the particle and calculate the direction in space from which the original neutrino must have come. It is highly ingenious. Moreover estimates suggest that DUMAND, if built to full size, ought to detect many celestial sources of neutrinos this way. In particular DUMAND ought to pick up the neutrino flux from supernovae going off in nearby galaxies as well as in our own. If neutrinos are massless they travel at light-speed, so the neutrino-burst will coincide with the light burst. But if they are heavy they will travel more slowly and the length of the measured delay will yield the neutrino-mass at last.

While experimental scientists are struggling to fund X-ray satellites, telescopes in space and huge detectors beneath the oceans, we must not forget that most fertile factory of scientific revolutions, the human brain. There is today an almost hysterical sense of anticipation amongst the small but brilliant group of theorists who work at the interface between particle physics and cosmology. There is in the air that rare but euphoric feeling that a momentous breakthrough between the two disciplines is about to take place. I recently attended a meeting held among some of them in a country house beside the Thames. Although I could not understand much of the abstruse jargon myself, I felt as if I was caught up with an international team of mountaineers who had just scaled an appalling overhang to find themselves at last within striking distance of the most challenging summit of all. To quote from a recent review in *Nature*:

Within the past 5 years there has been much exciting work and interesting developments in the inter-disciplinary fields of particle physics and cosmology. The primary tool of cosmologists has been the large telescope, but now they are interested in deep underground mine experiments built to detect proton decay, in experiments on mountain-tops searching for rare cosmic-ray events, in experiments performed in accelerators and reactors which are designed to detect axions and neutrino oscillations, in balloon flights looking for anti-protons, in high altitude U-2 flights which carry instruments to measure the anisotropy in the 3° microwave background, in low temperature searches for fractional charge states, in deep underwater muon and neutrino detectors, and even in a proposed monopole search at an iron-ore processing plant.

Traditionally, the large accelerator has been the primary tool of the particle physicist; however, they are now interested in reactor experiments, in deep underground mine experiments, in bulk matter searches for exotic particles left over from the big bang, in the cosmic abundance of helium and deuterium, in the distribution of galaxies, in the structure of the microwave background, and even in experiments designed to detect solar neutrinos. It is not uncommon to find as many cosmologists as particle physicists at a particle physics seminar, and vice versa. The connection between the two disciplines, is, of course, the big bang. In the early Universe temperatures as high as 10^{32} degrees Celsius were reached, corresponding to average particles energies as high as 10^{19} GeV. The particles present and their interactions determined the early evolution of the Universe, and the type of relics that remain. Particle physicists are constantly seeking higher energies where the world appears simpler, and the early Universe seems to be the only laboratory with large enough energies to study models of the unification of strong, weak and electromagnetic forces. Cosmologists are asking particle physicists if neutrinos have small rest masses, and if the proton is unstable. If neutrinos have small rest masses they may determine the large-scale dynamics of the Universe; and if the baryon number is unstable, baryon number isolating reactions may explain why the Universe is predominantly matter, rather than having equal amounts of matter and antimatter. Conversely, particle physicists are asking cosmologists about the cosmic environment impact of exotic particles which either cannot be produced at accelerators or would have escaped detection.

The seeds of any revolution thrive most vigorously amongst the tangled inadequacies of the present regime: so what deficiencies can we rec-

ognize already in the current big bang theory? First there is the homogeneity problem mentioned in Chapter 6. It looks as if, on the largest
observable scale, the Universe is extremely uniform, the same in one
direction as it is in any other. For instance, the cosmic background
radiation is uniform to one part in a thousand when, according to our
present understanding, it has no right to be. Two opposite poles of the
background that we can see now have never been causally connected,
that is to say no signal can ever have propagated between the two. How
then can they have been developed in unison? Second is the so-called
flatness problem, which is close to our own interest. Why is the Universe so nearly closed, why is Ω so close to 1 when in theory it could
have taken any value between nought and infinity? Third is the origin
of structure within the cosmos on the scale of galaxies and clusters.
Straightforward calculations show that any natural condensation of
such structures in the early radiation-dominated Universe was not permitted unless they had been built in by God or whoever since the very
beginning. Fourth is the gross disparity between the number of photons
and nucleons present: presently there are a billion photons for every
nucleon when once, so physics tells us, the numbers should have
been equal. Lastly there is the intellectual mist which swirls between
us and the earliest moments of Creation when conditions were too extreme for our current physical theories to apply. Beyond the mist all is
hidden.

The particle physicist faces difficulties of an even more daunting
nature. In the 1920s physicists invented quantum mechanics which
promised to explain the strange sub-atomic world in terms of a few
simple particles. Indeed success in atomic physics was almost totally
complete until men began probing deeper into the nucleus when they
found a bewildering variety of new phenomena and strange particles
like mesons and neutrinos for which there appeared no rhyme or reason. They found in addition four quite unrelated forces: gravity, electromagnetism, the 'weak force' mediating electrons and other leptons,
and the strong force governing baryons such as the proton and neutron.
As each new accelerator added its quota to the exotic zoo the theoreticians hobbled along behind looking for underlying patterns while adding strange theoretical conceptions of their own such as quarks and
gluons, colour and charm. They never gave up hope that beneath all
the chaos a more fundamental and unifying pattern would be found.
But they must have despaired at times. The difficult questions they
face are so numerous and so profound that it is difficult to list them in
any coherent way. They include: why are there four quite different

primary forces; why are there so many complex varieties of particle about when theoretically simpler conceptual entities like quarks and monopoles are never found in isolation? And why is there an asymmetry between the amounts of matter and antimatter in the Universe?

To describe why the log-jam seems to be on the move at last would probably require another book; certainly it would require another author. In any case the pattern is changing so fast at the moment, with each new theory being superseded within months by more sophisticated variants, and with claims and counterclaims being issued by experimentalists almost as frequently, that such a summary would be out of date before the ink was dry. All we can hope to do here is pick out a few highlights, skate over a glossary of buzz-words and prominent names, and mention some of the possible implications.

The excitement began about a decade ago when Abdus Salam from Imperial College, Stephen Weinberg from MIT and Sheldon Glashow from Harvard devised a new theory unifying the electromagnetic and weak forces. The theory predicted neutral currents which were identified soon afterwards at CERN (European Centre of Nuclear Physics). More experimental successes were to follow and it looks now as if CERN, in an experimental tour de force, has finally located the predicted W and Z bosons which are the particles supposed to mediate the unified electro-weak force.

In the successful Weinberg-Salam 'gauge' theory, as it is called, the vacuum is no longer empty but contains the so-called Higgs field (after another British physicist) which is responsible for the spontaneous appearance and decay of W and Z bosons. At high enough energies in the early Universe the electric and weak forces were identical, but as the cosmos cooled so their symmetry was broken, giving rise to the massive bosons so newly detected.

Because of its success the Weinberg-Salam theory has acted as a model for even more ambitious gauge theories called Grand Unified Theories (or GUTs for short) which attempt to unify the strong force with the electro-weak. Such theories are in a primitive state as yet, moreover they are difficult to check because their symmetries occur only at energies far too high for any conceivable accelerator, energies that would only have been attained in the furnace of the very early Universe. Hence the present dialogue between cosmologists and particle physicists. But one startling prediction is under study, with the results eagerly anticipated. The proton, hitherto regarded as indestructible, should, according to GUTs, decay with a half life of about 10^{31} years. Physicists are therefore crawling over huge heaps of scrap-iron buried deep in gold-mines looking for such rare decays.

Two applications of GUTs in particular excite the astrophysicist. Firstly GUTs allow for, though they do not definitely require, a finite mass for neutrinos. Secondly GUTs may explain why there is no significant antimatter in the Universe. According to the older theories particles and antiparticles should have been present in equal numbers and should all have annihilated each other by now, leaving a matter-less Universe. Yuri Sakharov, the brilliant Russian physicist now in exile in Siberia, has argued that when the cosmos cooled sufficiently for the GUT symmetries to break, then baryon-number-isolating reactions could have taken place which would leave a residue of particles after most particles and antiparticles had annihilated one another. This would account for the present matter left in the Universe.

One of the snags with GUTs is the embarrassingly large energies they attribute to the vacuum, energies sufficient to dominate the dynamics of the Universe and to close it many many times over. To overcome this difficulty Alan Guth from Stanford has resurrected a further term in the gravity equation, a term originally introduced by Einstein, but then dismissed by him as unnecessary and unaesthetic. This new term introduces a strong repulsion at long range and would have produced a very rapid expansionary or 'inflationary' phase in the early Universe. This newer 'inflationary' model of the Universe has several attractive features. During the rapid inflation, signals can propagate across what is now the whole observable cosmos, explaining the otherwise mysterious uniformity, while the observed presence of so many photons (10^9) for every nucleon, even the flatness of the Universe, may be accounted for by the latent heat of broken symmetries inherent in all GUTs.

The current simple version of GUTs are exciting more because they promise to tackle hitherto intractable problems like these than because of their detailed predictions, which in many cases go startlingly awry. For instance, one version predicts a Universe dominated by heavy magnetic monopoles which have never been seen for certain, while an inflationary Universe is one filled with galaxy-sized black holes, which is not the Universe we live in. Even so these theories have in them much room for detailed modification, leaving us with a brighter hope for the future than we ever had in the past. Their most satisfying feature, from a philosophical point of view, is that they may be able to generate the cosmos out of 'nothing' in a systematic way. Physicists have long been familiar with transitory or 'virtual' particles which are generated for brief intervals by random fluctuation in the vacuum: the more energetic the particle, the shorter its lifetime has to be, according to the famous Heisenberg Uncertainty Principle. Now gravitational energy is negative, mass energy positive, and the net balance between the two in the cosmos is observed to be close to zero. That is another way of saying Ω is close to 1. If the total energy is truly zero it is then possible to envisage the Universe as a long-lived fluctuation that has formed spontaneously, without prior cause, out of a pre-existing vacuum state. The need for a first cause, or a creator if you like, is thus removed one stage further back, though one could still look for the origin of the scientific laws themselves. Presumably if Universe can arise spontaneously it can also vanish one day with equal facility taking us all with it.

We are approaching the end of the book but not the end of the story. The next thirty years *could* be more exciting than the last. But I say 'could' rather than 'will' with some deliberation. Astronomy and particle physics have reached the point where new endeavours generally require the cooperative efforts of many people, that is to say they are expensive. For the past twenty years they have lived off the fat generated by the arms race and the space race. From now on explorers must rely on such funds as they can find for themselves, on such interest as they can generate among their fellow men.

I have to say the omens are not good. The world seems to be falling into the hands of dreary accountants who believe in the fallacy that if you do not spend money on X it automatically becomes available for Y. To them parsimony is a virtue. They stand on the dockside counting every yard of rope, asking self-importantly where it is all going to lead to, or where the short-term financial return is to be found. We can no more answer them than Columbus could, or Marco Polo, or all the other explorers who have ventured forth before or since. Science is not a business, it is human curiosity at play. It is not an investment, it is an entertainment. It is above all an incalculable gamble which, as we know, many societies have chosen not to take. The lights in the casino have gone out before; they could do so again. The curiosity of individuals can be taken for granted, but the curiosity of governments never can.

In Britain, the country where the communication satellite was invented if not developed, we spend each year on astronomy, space and radio research a paltry 40 million pounds, a mere 60 pence per man, woman and child. We can no longer launch scientific satellites of our own, nor even take the lead in starting international ventures. Our politicians presumably need the money to pay dole to people whose jobs they have taken away. We have opted out of the game. Our universities, where pure research largely originates, are so starved of funds that no jobs can be found in them for bright young people any more. After a few years of research they must leave to find jobs in a decaying industry, if they can. Astronomy is scarcely taught in schools, and only in tiny pockets on specialized courses in a few universities. Very few executive civil students and even fewer politicians have any training or interest in pure science at all. The Science Research Council, which was set up to fund pure science, has declined into the Science and Engineering Research Council with the emphasis moving towards applied research for short-term economic gain.

On the other side of the Atlantic prospects are brighter if only because the population is so much better educated than ours. The majority of Americans receive some scientific training at university level, and all the educated ones do. Even so the accountants and hard-nosed marketing men are moving in over there too and we see contraction and retreat on all sides. There are to be no more Americans on the moon this century, few really exciting ventures in space beyond the Space Telescope.

But I cannot believe man will turn his eyes from the stars for long. He came from the stars originally, and it is surely his destiny one day to return.

Notes on Technical Terms

Powers of ten

In astronomy very large cosmic numbers and very small atomic numbers are required. Since 10 x 100 = 1000, i.e. 10^1 x 10^2 = 10^3, this suggests the general rule that 10^a x 10^b = 10^{a+b}. Thus a very large number like a million, which is a thousand thousand, can be written 1000 x 1000 = 10^3 x 10^3 = 10^{3+3} = 10^6, which is a compact and clear notation indicating that a million is 6 tens multiplied together. Likewise since 100 x $\frac{1}{10}$ = 10, i.e. 10^2 x $\frac{1}{10}$ = 10^1, this suggests we write $\frac{1}{10}$ as 10^{-1}, for then, according to the rule 10^a x 10^b = 10^{a+b}, we have 10^2 x 10^{-1} = $10^{2+(-1)}$ = 10^1 = 10, which is correct. A millionth, which is 1 over six tens multiplied together, can be written as 10^{-6}. Thus the number of atoms in the visible Universe, a very large number, can be written compactly as 10^{80}, whilst the very short time it takes light to cross the nucleus of an atom can be written compactly as 10^{-23} secs.

The Hubble constant H_0

The so-called Hubble constant is a measure of how fast the Universe is expanding. H_0 is actually the speed of recession of a galaxy 3 million light years away, and is roughly 100 km/sec. A galaxy twice as far (6 million l.y.) will be expanding at $2H_0$ or 200 km/sec . . . and so on. If you think about it, $1/H_0$ is the time since the two galaxies were piled right on top of us, i.e. it is the time since the Big Bang, sometimes called 'the age of the Universe.' The Hubble constant does not change in space but can vary slowly with time.

The cosmic density parameter Ω

According to Einstein the geometry, the finiteness and the future fate of the Universe all depend on its density. If the present density ρ exceeds a critical density ρ_c (= 1 atom/cubic metre at present) then the Universe is finite, and will recollapse. If ρ is less than ρ_c it is infinite and will expand for ever. Thus the ratio of actual density ρ to the critical density ρ_c, which we call Ω, i.e. $\Omega = \rho/\rho_c$, is of great importance in astronomy. If Ω is more than 1 the Universe is finite and will recollapse, if less than 1 it is infinite and will expand for ever. What matters is not the precise value of Ω, which can vary somewhat with time, but whether Ω is greater or less than 1.

The mass to light ratio Γ

This ratio is a measure of how much mass or weight we are to associate with so much visible light from a given object. Thus the sun weighs 2500 tons for every kilowatt of heat and light it emits, and we arbitrarily ascribe to it a mass to light ratio of Γ(gamma) = 1. A star which weighs 5000 tons/kilowatt of emitted energy then has a Γ of 2. Individual stars generally have Γs in the range 0.1 to 400. A mixed population of normal stars will have an overall Γ in the range of 2 to 10. Because individual galaxies contain much invisible non-luminous matter they have much higher Γs around 50. If the Universe is closed, i.e. if Ω is more than 1, it will be necessary for the average value of Γ for all the material in the Universe to have a value around 1000.

The gravitational constant G

This constant is a measure of the strength of gravitation. The value of G can be measured by delicate experiments in a laboratory where one measures the minute pull or force exerted by two heavy balls upon one another. By the equations in Chapter 5, that force is $F = Gm_1m_2/r_{12}^2$, where m_1 and m_2 are the masses of the two balls and r_{12} the distance between their centres. After measuring F, m_1, m_2 and r_{12}, it is then possible to calculate G as $= Fr_{12}^2/m_1m_2$ once and for all, and that constant value (about 6.7×10^{-8} in centimetre, gm, second units) can be used in all subsequent calculations. Gravitation is in fact very weak. For instance the gravitational force exerted by two space-walking astronauts on each other is so feeble that even if they were put 1 metre apart their mutual gravitation would need to act for 3 hours to pull them together.

$E = mc^2$

Einstein showed in 1905 that unless an amount of energy E possesses a minute but definite amount of mass $m = E/c^2$, where c is the velocity of light, you have paradoxes. For instance he showed that you would be able to propel a spacecraft indefinitely simply by repeatedly exchanging a hot and cold but otherwise identical pair of glasses of water from end to end of the craft. Attributing to the hot, and therefore slightly more energetic (by an amount E) glass, an extra mass $m = E/c^2$ removes this paradox.

The law $m = E/c^2$ finds two main applications in astronomy. Rearranged as

$E = mc^2$ it tells us that if mass can be somehow converted into energy, as happens in the thermonuclear reactions, within stars a vast amount of energy will result because the speed of light c, appearing in the equations, is such a large number, i.e. 3×10^{10}cm/sec. Secondly it tells us that energy itself has mass and therefore exerts a gravitational pull. In fact very massive stars may be destabilized by the self-gravitating effect of their own internal radiation.

eV

An eV or *'electron-volt'* is a measure of energy and/or mass used on the atomic scale. An electron, the lightest particle known to have a definite mass, weighs about half a million eV. Neutrinos, if they weigh anything at all, have masses of less than 100 eV. But they are so numerous that even if they weigh as little as 30 eV each, they could in sum dominate the mass of the Universe.

Bibliography

This is not intended to be a complete survey of what has become a vast technical literature. Wherever possible the references are to reasonably accessible articles which demand of the reader no more expertise than is required to read this book. The more technical references (marked with an asterisk) will be difficult to find outside a university library. The key to abbreviations is at the end.

Chapters 1 and 2
Frontiers of Astronomy. Hoyle, F., Heinemann.
Discovery in Astronomy. Harwit, M., Harvester Press, 1981.
The Stars, Their Structure and Evolution. Tayler, R., Wykeham Science, 1974.

Chapters 3 and 4
The Milky Way. B. and P. Bok, Harvard Univ. Press, 1981.
'The Milky Way Galaxy.' Bok, B., *Sc.Am.*, March 1981, 70.
The Discovery of Our Galaxy. Whitney, C., Angus and Robertson, 1973.
'The Andromeda Galaxy.' Hodge, P., *Sc.Am.*, Jan. 81, 88.
The Realm of the Nebulae. Hubble, E., Yale Univ. Press, 1981.
'The Ultimate Fate of the Universe.' Islam, J., *QJRAS*, 18, 1977.
'The Clustering of Galaxies.' Groth, E. et al., *Sc.Am.*, Nov. 1977, 76.
'The Future of the Universe.' Dicus, D. et al., *Sc.Am.*, March 1983, 74.
Man Discovers the Galaxies. Berendzen, R., Hart, R., Sealey, D., Sci. History Publs., 1976.
*Dyson, F., 1979, *Rev.Mod.Phys.*, 51, 447 (fate of Universe).

Chapters 5, 6 and 7

The First Three Minutes. Weinberg, S., André Deutsch, 1977.

Cosmology. Harrison, E., Cambridge Univ. Press, 1981.

The Big Bang. Silk, J., Freeman, 1980.

Relativity. Einstein, A.

Gravity. Gamow, G., Doubleday.

'Newton's Apple and Galileo's Dialogue.' Drake, S., Sc.Am., Aug. 1980, 122.

'The Structure of the Early Universe.' Barrow, J., Silk, J., Sc.Am., April 1983, 98.

*Essential Relativity. Rindler, W., van Nostrand.

*Tinsley, B., 1977, Ap.J., 173, L93 (galaxy evolution).

*Ostriker, J., Hausman, M., 1977, Ap.J., 217, L125 (galaxy cannibalism).

*Schramm, D., Waggoner, R., 1977, Ann.Rev.Nucl. and Part.Sci., 27, 37 (deuterium).

'Will the Universe Expand Forever?' Gott, J. et al., Sc.Am., March 1976, 62.

'The Cosmic Background Radiation and the New Aether Drift.' Muller, R., Sc.Am., May 1978, 64.

*Sandage, A., Tammann, G., Hardy, E., 1972, Ap.J., 172, 253 (Ω(dyn)).

'The Search for Black Holes.' Thorne, K., Sc.Am., Dec. 1974, 32.

*Shapiro, S., Astron.J., 1971, 76, 291 (Ω(vis)).

*Zang et al., 1982, QJRAS, 23, 363 (Ω).

Chapters 8 and 9

*'Masses and Mass to Light Ratios of Galaxies.' Faber, S., Gallagher, J., 1979. Ann.Rev.Astron.Astrophys., 17, 135.

*Distribution and Kinematics of Neutral Hydrogen in Spiral Galaxies of Various Morphological Types. Bosma, A., 1978, Ph.D. thesis, Univ. of Groningen.

'Dark Matter in Spiral Galaxies.' Rubin, V., Sc.Am., June 1983, 88.

*Page, T., 1952. Ap.J., 116, 63. (double galaxies).

*Turner, E., 1976, Ap.J., 208, 304 (double galaxies).

*Fabricant, D. et al., 1980, Ap.J., 241, 552 (M87).

*Luyten, W., 1968, MNRAS, 139, 221.

*Turner, E., Aarseth, S. et al., 1979, Ap.J., 228, 684.

*Miller, R. et al., 1970, Ap.J., 181, 903 (computer model galaxies).

*Ostriker, J., Peebles, P., 1973, Ap.J., 186, 467 (computer model galaxies).

*Mattilla, K., 1976, A. and A., 44, 77 (EBL).

*Spinrad, H. and Stone, R., 1978, Ap.J., 226, 609 (EBL).

*Dube et al., 1979, Ap.J., 232, 327 (EBL).

*Zwicky, F., 1937, Ap.J., 86, 217.

Chapter 10

The Search for Gravity Waves. Davies, P., 1980, Cambridge Univ. Press.

*Carr, B. J., 1979, A. and A., 89, 6 (gravity waves).

'Neutrinos.' Morrison, P., *Sc.Am.*, Jan. 1956, 58.
'Cowsik, R. and McClelland, J., 1973, *Ap.J.*, *180*, 7 (neutrinos).
*'Neutrinos in the Universe.' Tayler, R. 1981, *QJRAS*, *22*, 93.
'The Quantum Mechanics of Black Holes.' Hawking, S., *Sc.Am.*, Jan. 1977, 34.
'The Einstein X-ray Observatory.' Giacconi, R., *Sc.Am.*, Feb. 1980, 70.
*Carr, B. J., 1978, Comments in Astrophysics, 7, 161 (black holes).

Chapter 11
*Reid, I., Gilmore, G., 1982, *MNRAS*, *201*, 73 (red dwarfs).
*Schmidt, M., 1975, *Ap.J.*, *202*, 22 (red dwarfs).
'The Companions of Sun-like Stars.' Abt, H., *Sc.Am.*, April 1977, 96 (dwarfs).
*van Biesebroek, 1961, *AJ*, *66*, 528 (black dwarfs).
*Rees, M. J., 1976, *MNRAS*, *176*, 483 (black dwarfs).
*White, S., Sharp, N., 1977, *Nature*, *269*, 395 (stickiness).
*Tremaine, S., Gunn, J., 1979, *Phys.Rev.Letts.*, 42 E 407 (neutrinos).
Schramm, D., *Physics Today*, April 1983, 36, No. 4, 27 (neutrinos).
*Bond, J. and Carr, B., 1984, *MNRAS*, *207*, 585 (black holes).
'Origin of the Cosmic X-ray Background.' Margon, B., *Sc.Am.*, Jan. 1983, 94.
*Field, G., Perrenod, S., 1977, *Ap.J.*, *215*, 717 (IGG).
Silk, J., 1981, *Nature*, *280*, 83 (IGG).
*Bean, A. et al., 1983, *MNRAS*, *205*, 605 (Ω(dyn)).

Chapter 12
Iceberg theory
*Fish, R. A., 1964, *Ap.J.*, *139*, 284.
*Freeman, K. C., 1970, *Ap.J.*, *160*, 811.
*Disney, M., 1976, *Nature*, *263*, 573.
*Disney, M., Phillipps, S., 1983, *MNRAS*, *205*, 1253.

Unconventional gravitational theories
*Rood, H., 1974, *Ap.J.*, *193*, 15.
*Finzi, A., 1963, *MNRAS*, *127*, 21.
*Jackson, J. C., 1970, *MNRAS*, *148*, 249.
*Stacey, F., Tuck, G., *Nature*, *292*, 230.
*Milgrom, M., 1983, *Ap.J.*, *270*, 365.

Chapter 13
*First Results from IRAS 1984, *Ap.J.*, *278*, No. 1, Pt. 2 (whole vol.).
'Charge Coupled Devices in Astronomy.' Kristian, J., Blouke, M., *Sc.Am.*, Oct. 1982, 48.
'The Space Telescope.' Bahcall, J., Spitzer, L., *Sc.Am.*, July 1982, 38.
'The Early Universe and High Energy Physics.' Schramm, D. N., *Physics Today*, April 1983, 36.
'The Early Universe.' Kolb, E., Turner, M., 1981, *Nature*, *294*, 521.

'A Deep-Sea Neutrino Telescope.' Learned, J., Eichler, D., Sc.Am., Feb. 1981, 104.

'A Unified Theory of Elementary Particles and Forces.' Georgi, H., Sc.Am., Apr. 1981, 40.

'Beyond the Big Bang.' MacRobert, A., Sky and Telescope, March 1983, 211.

'The Inflationary Universe.' Guth, A., Steinhardt, P., Sc.Am., May 1984, 90.

The abbreviations stand for the names of the following journals:

QJRAS   Quarterly Journal of the Royal Astronomical Society
Sc.Am.   Scientific American
Ap.J.   Astrophysical Journal
MNRAS   Monthly Notices of the Royal Astronomical Society
AJ   Astronomical Journal
Phys.Rev.Letts   Physical Review Letters
Rev.Mod.Phys.   Reviews of Modern Physics
A. and A.   Astronomy and Astrophysics

Index